Geometric Reasoning

Edited by

John Woodwark

IBM UK Scientific Centre
Winchester

CLARENDON PRESS · OXFORD
1989

Oxford University Press, Walton Street, Oxford OX2 6DP

Oxford New York Toronto
Delhi Bombay Calcutta Madras Karachi
Petaling Jaya Singapore Hong Kong Tokyo
Nairobi Dar es Salaam Cape Town
Melbourne Auckland
and associated companies in
Berlin Ibadan

Oxford is a trade mark of Oxford University Press

Published in the United States
by Oxford University Press, New York

British Library Cataloguing in Publication Data
Geometric reasoning.
1. Computer systems. Graphic displays. Programming.
Applications of geometry
I. Woodwark, John
006.6'6
ISBN 0-19-853738-7

Library of Congress Cataloging in Publication Data
(Data available)

Printed in Great Britain by
Courier International Ltd
Tiptree, Essex

CONTENTS

Contents

15   Using CAD and expert systems for human workplace design   269

Maurice Bonney, Nick Taylor, and Keith Case

Participants

Dr B.S. Acar	Department of Engineering Production, University of Loughborough, Loughborough LE11 3TU.
Mr C. Attwood	University of Reading, Whiteknights, Reading RG6 2AB.
Prof M.C. Bonney	Department of Production Engineering and Production Management, University of Nottingham, University Park, Nottingham NG7 2RD.
Dr A. Bowyer	School of Mechanical Engineering, University of Bath, Claverton Down, Bath BA2 7AY.
Dr I.C. Braid	Three-Space Ltd, 70 Castle Street, Cambridge CB3 0AJ.
Mr A. Buchanan	Geometric Modelling Project, Department of Mechanical Engineering, Leeds LS2 9JT.
Dr S.A. Cameron	Department of Engineering Science, University of Oxford, Parks Road, Oxford OX1 3PJ.
Miss L.W. Chan	Cambridge University Engineering Department, Trumpington Street, Cambridge CB2 1PZ.
M H. Crapo	INRIA, BP 105, 78153 Le Chesnay, Cedex, FRANCE.
Dr R.N. Cuff	IBM UK Scientific Centre, St Clement Street, Winchester SO23 9DR.
Prof J.H. Davenport	School of Mathematics, University of Bath, Claverton Down, Bath BA2 7AY.
Mr J. Dodsworth	Geometric Modelling Project, Department of Mechanical Engineering, Leeds LS2 9JT.
Dr R.A. Earnshaw	Department of Computer Studies, University of Leeds, Leeds LS2 9JT.
Prof F. Fallside	Cambridge University Engineering Department, Trumpington Street, Cambridge CB2 1PZ.
Mr R.B. Fisher	Department of Artificial Intelligence, University of Edinburgh, Forest Hill, Edinburgh EH1 2QL.
Mr A.D. Fleming	Department of Artificial Intelligence, University of Edinburgh, Forest Hill, Edinburgh EH1 2QL.
Prof A.R. Forrest	School of Information Systems, University of East Anglia, Norwich NR4 7JT.

Dr A. Geisow   Hewlett-Packard Laboratories, Filton Road, Stoke Gifford, Bristol BS12 6QZ.

Prof S.P. Grau   School of Industrial Engineering, Universidad Politechnica de Valencia, Apartado 22012, Valencia, SPAIN.

Dr M.A. Gray   IBM UK Scientific Centre, St Clement Street, Winchester SO23 9DR.

Mr A.R. Halbert   IBM UK Scientific Centre, St Clement Street, Winchester SO23 9DR.

Dr T.R. Heywood   IBM UK Scientific Centre, St Clement Street, Winchester SO23 9DR.

Dr D.H. Irvine   Polytechnic of Central London 115 New Cavendish Street, London W1M 8JS.

Mr G.E.M. Jared   Department of Mathematics, Cranfield Institute of Technology, Cranfield MK43 0AL.

Dr J.M. Knapman   IBM UK Scientific Centre, St Clement Street, Winchester SO23 9DR.

Miss D. Lee   IBM UK Scientific Centre, St Clement Street, Winchester SO23 9DR.

Dr R.R. Martin   Department of Computing Mathematics, University College Cardiff, Mathematical Institute, Senghennydd Road, Cardiff CF2 4AG.

Dr A.E. Middleditch   Polytechnic of Central London 115 New Cavendish Street, London W1M 8JS.

Mr F.G. Mill   Department of Mechanical Engineering, University of Edinburgh, The King's Buildings, Edinburgh EH9 3JL.

Mr P.S. Milne   School of Mathematics, University of Bath, Claverton Down, Bath BA2 7AY.

Prof J.L. Murray   Department of Mechanical Engineering, Heriot-Watt University, James Nasmyth Building, Piccarton, Edinburgh EH14 4AS.

Dr M.J.L. Orr   Department of Artificial Intelligence, University of Edinburgh, Forest Hill, Edinburgh EH1 2OL.

Mr R.D. Parslow   Parslow Associates Ltd, 6 Ormond Avenue, Hampton-on-Thames, Middlesex TW12 2RU.

Mr N.R. Phelan   School of Mechanical Engineering, University of Bath, Claverton Down, Bath BA2 7AY.

Mr J. Porrill   AI Vision Research Unit, University of Sheffield, Sheffield S10 2TN.

Mr D. Prior   Department of Artificial Intelligence, University of Edinburgh, Forest Hill, Edinburgh EH1 2QL.

Dr P. Quarendon IBM UK Scientific Centre, St Clement Street, Winchester SO23 9DR.

Dr K.M. Quinlan Computervision European Development Group, Computervision House, Penn Street, Amersham HP7 0PX.

Mr S.T. Rake IBM UK Scientific Centre, St Clement Street, Winchester SO23 9DR.

Mr R.J. Richards Cambridge University Engineering Department, Trumpington Street, Cambridge CB2 1PZ.

Dr M.A. Sabin FEGS Ltd, Oakington, Cambridge CB4 5BA.

Mr R.K. Stobart Cambridge Consultants Ltd, Science Park, Milton Road, Cambridge CB4 4DW.

Mr C. Suffell Department of Combined Engineering, Coventry (Lanchester) Polytechnic, Priory Street, Coventry CV1 3FB.

Mr N.K. Taylor Department of Production Engineering and Production Management, University of Nottingham, University Park, Nottingham NG7 2RD.

Mr S.J.P. Todd IBM UK Scientific Centre, St Clement Street, Winchester SO23 9DR.

Mr A.F. Wallis School of Mechanical Engineering, University of Bath, Claverton Down, Bath BA2 7AY.

Dr D.J. Williams Manufacturing Engineering Group, Cambridge University Engineering Department, Mill Lane, Cambridge CB2 1PZ.

Prof M.H. Williams Department of Computer Science, Heriot-Watt University, Piccarton, Edinburgh EH14 4AS.

Dr J.R. Woodwark IBM UK Scientific Centre, St Clement Street, Winchester SO23 9DR.

Mr S. Wright Cambridge University Engineering Department, Trumpington Street, Cambridge CB2 1PZ.

Introduction

It is trite to observe that a computer can do all the multiplications that the average human does in a lifetime in a fraction of a second but that a programme exhibiting 'common sense' has yet to be written. Anything that can be reduced to a simple algorithm can be done easily by a computer; other things are very difficult either because algorithms are difficult to formulate or have intrinsically rotten performance. 'Expert systems' are perhaps the best-known attempt to get around this problem, and the difficulties in constructing them and their apparent limitations are also widely discussed.

When computers are asked to deal with information about the shape of objects, the same separation of tasks into those that can be attacked by algorithms and those that cannot is observed. In computer graphics, for instance, it was perhaps always apparent that the mechanics of picture synthesis would be amenable to algorithmic treatment, to the extent that it is a mechanical simulation of the physics of light. Even so, topics such as hidden-surface elimination dominated the graphics literature for ten years and more. In image analysis, it was perhaps less obvious that techniques essentially taken from signal processing could be applied in two dimensions to give a range of low-level image-processing primitives that would actually be useful in clarifying an image; but such was the case.

While computer graphics and image processing have been with us for thirty years or so, and look like mature disciplines, it is really only the algorithmic processes that have been computerized. There are many more complex, less easily pinned-down things that we still want to do with shape information and cannot. In the organization of this conference, we called these activities **Geometric Reasoning**.

These Proceedings are essentially a snapshot of UK activity on topics coming under this Geometric Reasoning umbrella. The UK focus was only achieved by not publicizing the event abroad and, in the end, we did include a paper generously offered by our colleagues across the Channel. In content, rather than geographical origin, the papers are very wide-ranging, spanning many different applications of geometry in computing, and reporting work with many different goals, from the very theoretical to the very practical. We have certainly not restricted ourselves to the topic of **computational geometry**, which often seems to offer theoretical bounds and worst-case analyses that would be utterly devastating if we did not deploy theoretically less elegant techniques. Nor do we look only at classical von Neumann computers, as a paper describing the use of a connection machine is included.

In terms of applications, the spread is equally wide, encompassing shape modelling, computer graphics, image processing and computer vision. Every author has something to add to the subject, none has a complete answer. In many areas, such as feature recognition for process planning and industrial machine vision, enormous commercial benefits are potentially at stake. Progress is being made, but the problems are much less tractable than those that we solved ten years ago, when a single algorithm or approach would suddenly revolutionize an aspect of image processing or graphics. Expect progress to be slow and steady at best; many of the problems discussed in this book will still be with us in the next millennium.

John Woodwark
IBM UK Scientific Centre, Winchester

1 A geometric algebra system

A. BOWYER, J. H. DAVENPORT, P. MILNE, J. PADGET and
A. F. WALLIS

1.1 Introduction

Existing computer algebra systems are not particularly well suited to the solution of geometrical problems. This is not because the underlying theory suffers from any inherent limitation, but because the need to provide algorithms and systems for the solution of such problems has not been addressed by the computer algebra research community.

The authors will therefore be constructing a **geometric algebra system**. This will be a computer algebra system biased towards the solution of geometric problems mainly, but not exclusively, of the types encountered in solid modelling. It will be written in the form of a library, so that its users may employ it to solve algebraic problems in the same way as, for example, people now write software that obtains numerical answers to problems by calling the NAG library.

Among the facilities being implemented are general integer, rational number and polynomial arithmetic (but oriented towards the perceived needs), resultant calculations, real root finding, and the types of substitutions needed to manipulate geometric objects.

We discuss first the domains that we wish the modellers to have, and then the algebraic facilities required. This leads into a discussion of algebraic data structures, followed naturally by our solution to the thorny problem of computer storage management.

1.2 Model domains

The first technical question that someone who is about to write a solid modelling system asks is 'What objects do I wish the system to be able to represent?' The answer to this question is known as the **model domain** of the system. A model domain has many independent aspects. Here are a few:

- Surface complexity: the modeller may only be able to represent objects with flat faces, or it may be able to represent objects with a wide variety of curved surfaces. This will be discussed more fully in the next section.

- Query domain: this will be discussed in more detail below as well. Essentially this is the range of questions about modelled objects which the modeller is able to answer.

- Model order: The modeller may only be able to represent objects with a few instances of primitive shapes, say up to one hundred, or it may be able to accommodate thousands. This will clearly be decided, in part, by the memory capacity of the computers on which the modeller is being implemented. But more important considerations are the inherent compactness of its data structures and the efficiency of the modeller's algorithms against model order. (For example, if there are n instances of primitive shapes, does the modeller take a time $O(n^2)$ to produce a picture of the model, or a time $O(n \log n)$?) Memory space is, more or less, an absolute constraint (even in virtual systems). What constitutes an acceptable computation time depends upon the patience of the user: a commodity commonly held by all but system vendors to be in short supply.

- Non-geometrical information: In addition to recording the shape of an object, a modelling system may also record information about what it is made from, by what process each of its component parts is to be manufactured, how much each object costs, and so on. Whilst clearly very useful, this sort of information is beyond the scope of this paper, and will not be mentioned again.

Until fairly recently, many commercial solid modellers could represent only faceted objects (curved surfaces being represented by polyhedral approximation), and exhibited markedly worse-than-linear computation times against model order. Notable exceptions are modellers such as DUCT [21], a boundary modeller which allows its users to construct objects using parametric curved patches, and Synthavision, an early set-theoretic modeller which has highly inefficient evaluation algorithms. The authors and other researchers [10,11,14,29,31,32,33] have also, in the last couple of years, developed very efficient set-theoretic solid modellers that have a primitive domain consisting of, essentially, any surface that may be described by an implicit polynomial in the three independent space variables (x, y, z). These modellers exhibit a very wide model domain, which is clearly a good thing from their user's point

Fig. 1.1 An archeological reconstruction of part of the Roman settlement at Caerleon in Wales produced using the DORA modeller from the University of Bath.

of view as long as the method provided for the user to describe his complicated model is easy to employ. Model input is discussed below.

Primitive domains

As was briefly mentioned above, the primitive domain of a modeller is the range of surface types that the objects modelled with it can have.

The simplest primitive domain that a modeller may have is that which allows it to hold only faceted models. That is, the domain contains a single primitive: a flat plane. The reconstruction of the Roman Baths at Caerleon shown in Fig. 1.1 was generated by the DORA modeller from the University of Bath, which is a modeller of this kind. This type of modeller will serve adequately in architecture, for instance, as most buildings really are faceted. It will, perhaps surprisingly, go a long way towards meeting the needs of mechanical engineering as well, despite the fact that all curved surfaces have to be approximated. The approximation can be controlled to lie within acceptable tolerances, so that calculations of, for example, the mass of the object modelled, or the generation of a finite element mesh inside it, can be carried out with confidence. Faceted models are not really suitable, however, for the generation of cutter paths for the automatic production of the objects that the models represent, or for certain types of simulation, such as the calculation of fluid flow, or the operation of mechanisms.

Another very simple primitive domain with a single primitive is that spanned by spheres modellers. Here objects are approximated by the union of a large number of small spheres (sometimes, but not necessarily, all the same diameter). Spheres models are very easy to render. All that is needed is a depth sort of the spheres in distance from the eye. They do, however, necessarily have lumpy surfaces, which makes them unsuitable for many engineering applications. Their forte is, of course, in molecular modelling.

Most authors of faceted modellers have subsequently extended their primitive domain by adding useful quadrics and the torus to the original flat primitive. (In passing, it is worth mentioning that many faceted modellers do not have an infinite plane as their only primitive, but instead employ the cuboid and, sometimes, the wedge. These have the advantage of spatial locality, so it is easy to keep track of them using boxing tests.) The phrase 'useful quadrics' can, in almost all circumstances, be taken to mean the cylinder, cone and sphere. A modeller with this primitive domain can represent a very large number of engineering components accurately. It cannot, however, represent components composed of these shapes with fillets added between them, except when such a fillet would have the form of one of the primitives (a toroidal fillet, for instance, would suffice at the junction between a plane and a cylinder with its axis perpendicular to the plane, but not when the axis was inclined). This makes such models less useful in many fields (such as pattern-making and injection moulding; fillets on dies and moulds are essential for their successful operation).

Plane-quadric-torus modellers do not allow models such as that depicted in Fig. 1.2 to be constructed. As can be seen, this model has a large number of swept free-form curves and smooth fillets. It was created by the latest modeller written at the University of Bath [29]. This modeller allows its user to define shapes using any implicit polynomial that he or she chooses up to quite high degrees (16 or so). Using the shapes that this freedom allows, virtually any engineering component may be accurately modelled.

Sadly, as in the rest of life, freedom is never free. The cost that has to be paid for the versatility of the latest Bath modeller is algebraic complexity. A polynomial of total degree 16 in the three independent space variables x, y and z has 969 coefficients. The way that the polynomials are defined is described in the Appendix. Finding their roots for picture creation [29], or for cutter path generation [32], can be done quite successfully using classical numerical techniques, but starts to give problems when the degrees of the polynomials get large.

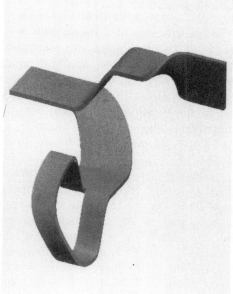

Fig.1.2 A model from the latest Bath set-theoretic modeller.

Fig.1.3 Twists modelled by the gap-blend technique described in the Appendix.

Finally, in addition to simple polynomials, it would be possible for a modeller to admit rational functions (quotients of two polynomials) to its primitive domain, and, ultimately, transcendental functions as well. This last addition is not as frivolous as it might at first seem. It is often important to be able to model surfaces which, potentially, have an infinite number of roots. The helix is the obvious example, and a clearly useful one from an engineering point of view. We do not propose to address these functions in this paper.

Obviously, the more complicated the primitive domain a modeller has, the more versatile it is. Also, the more complicated the domain becomes, the more subtle are the algebraic techniques that the modeller needs to employ to evaluate the model efficiently. The research project for which this paper forms an initial report is directed towards the goal of incorporating symbolic algebra techniques into solid modelling to make a geometric algebra system.

The work reported here is concerned with the application of symbolic algebra to set-theoretic modellers, and so this section has concentrated on their primitive domains. There are, of course, a roughly parallel set of primitive domains from facets to parametric patches that are available in boundary modellers of various degrees of sophistication.

Query domains

The construction of a solid model would be entirely pointless if it were not possible subsequently to ask questions about it. The user may wish to know the volume of the object modelled, for instance, or a rendering program may wish to know which surface of the object a ray from its ray-caster strikes first, and the surface normal at the struck point. The range of questions, generated manually or automatically, which the modeller is able to answer unambiguously, accurately and efficiently is known as its **query domain**. In principle, the larger this domain the better.

Clearly, many complicated queries are also implicitly composite. That is to say, for example, the query 'What does the object modelled look like?' can be re-posed as a large number (in this case) of simpler queries such as the ray-casting one just mentioned. Queries, then, can be considered to be either 'fundamental' (that is to say they are irreducible, and must be answered directly by the modeller) or 'composite'. The query domain is really the set of fundamental queries. Once these have been supported composite queries can be coded in terms of them until the author of the system is satisfied (or, of course, until the user is satisfied: usually a more distant milestone).

Perhaps the most fundamental query of all is the 'membership test'. Given a point in space (x,y,z) and a solid model, a membership test returns a logical value indicating whether the point is inside or outside the object modelled (numerical accuracy and fudge factors permitting, it may also indicate if the point is on the object's surface). It is possible, for instance, to code the composite query 'What is the volume of the object?' using a Monte Carlo technique making repeated calls to the fundamental membership test. This also allows masses and moments of inertia to be calculated. By slightly modifying the technique, membership tests also allow the calculation of surface areas [23].

The fundamental ray-casting query has already been mentioned. In addition to picture rendering, it is also useful in tool path generation [32], and in taking measurements from pictures of a model interactively (useful for checking) [19, 20].

Simple-sounding queries tend to be composite, whereas fundamental queries tend to sound abstract. An example of a simple-sounding query which is, in fact, often hard to answer is 'Where is (part of) the model?' Sometimes the ray-casting query can be used to answer this, but often a spatial search in more than one dimension is needed. In order to perform this efficiently divide-and-conquer techniques can be used (see Section 1.4). These need answers to fundamental queries such as 'Does this box lie inside, lie outside, or cut the surface of the object?' If the object is composed of complicated polynomial primitives this is not an easy question to answer, though the authors and colleagues have worked on efficient numerical solutions to this problem [29].

Once again, the authors intend to use algebraic rather than numerical techniques in order to improve the range, efficiency, and accuracy of set-theoretic solid modelling query domains.

1.3 User input

As described in the Appendix, it is not necessary for the user of a modelling system which supports any polynomial surface as its primitive domain to describe the polynomials that are to be used by computing their coefficients; they can be built naturally from simple parts of the model. Model definitions can be built using a model input language. Alternatively, some systems allow their users to sketch the objects that they want on a graphics screen, and the result is tidied up and built into a model. The most powerful and useful systems allow their users to combine these techniques, as they both have strengths and weaknesses.

The authors use a Pascal-like language called SID [2,23] for model definition. SID is an acronym for Set-theoretic Input to DORA (a previous faceted modeller written at Bath). SID allows its users complete freedom to build polynomially-surfaced objects by using the techniques described in the Appendix. It has a rich set of built-in functions and operators that allow its user to do vector algebra to facilitate geometric calculation, and to construct parameterized models as functions which can be called. It also allows its user to read models from other sources (such as a sketching program of the type referred to above), to manipulate them, and to include them in larger models. The language is recursive, and so a simple program can be used to generate quite complicated objects.

Generating faceted models using sketch input is straightforward, but a subtle decomposition algorithm [30] is needed to turn such input into a canonical set-theoretic form. Most sketch input programs allow their users to extrude a sketched outline to form a solid (possibly with a draft angle to form the frustum of a cone) or to sweep their sketch about an arbitrary axis to form a solid of revolution. In principle it would be possible to sweep an outline along any space curve to form a solid, and the authors will be investigating the automation of the algebra needed to do this. Outlines that include curved sections are more problematic, as the decomposition algorithm mentioned above needs to be able to compute the convex hull of the sketched outline. This is a well-understood problem if it is a polygon, but not so straightforward if it contains polynomial sections. The authors will be investigating the algebra needed to solve this problem as well.

Once the model has been created, it is obviously useful if it can be edited interactively, especially to tweak the numerical values (particularly those used to define blends). This may be done using a program to generate cross-sections of models quickly.

1.4   Division strategies: pruning

Object-space division and model pruning have been used in all the solid modelling systems written at the University of Bath during the last six years in order to improve their efficiency radically.

This divide-and-conquer technique recursively divides the volume containing the model, thus generating new sub-volumes; a new sub-model is generated for each sub-volume by pruning the model for its 'father' sub-volume. The sub-volumes are stored in a tree structure, with the leaf volumes containing sub-models. Pruning the model for a sub-volume is achieved by comparing each instantiated primitive of the model with the sub-volume and replacing those primitives that do not intersect the sub-volume with references to air or solid. The sub-model may then be simplified by applying the definition rules for OR (union) and AND (intersection) to its set-theoretic definition (for example, anything intersected with air may be replaced with air).

The effect of this technique is to localize the geometric complexity present in the original model. Downstream processes (for example the generation of a picture) that use the model may then utilize these localized models instead of the original. Details of the technique, which has been used in the generation of both shaded and wire-frame pictures are given in [23,24,25,28].

At each stage in the division process a sub-volume may be split into two or more new sub-volumes. It is not clear that division into more than two holds any real advantages over simple binary division. Whatever the type of division, the form, position and orientation of the dividing surface(s) has to be chosen.

The simplest alternative is to divide alternately in each of the coordinate directions using a plane positioned so that it passes through the centre of the sub-volume. This generates an isotropically divided structure of cuboids with their faces aligned with the coordinate system. A more complicated alternative is to choose the axis orientation and position of the division plane by applying a geometric tests to the sub-model. This is the technique used by the zone pruner in DORA [23].

Such a division plane may also be given an arbitrary orientation. This potentially allows the model to be pruned more effectively in relation to the divided space and can hence result in a smaller divided structure for any given level of division. An even more involved alternative would be to use a non-planar division surface, which could result in an even smaller structure.

Whatever the division strategy used, a number of decisions have to be made during the division process: whether or not to divide a sub-volume further and if so, where to divide and using what type of surface. The decision whether or not to divide a sub-volume may be made using simple non-geometric tests based on the sizes of the sub-model and sub-volume. However, a decision based on testing the effect of further division on the divided structure will generate a more efficiently divided structure. In this case the decision will also be influenced by the choice of location and orientation of the division surface.

If a more efficiently divided structure is to be generated by calculating the position of the division surface, this can only be made by knowing the effect that further division will have on the downstream processes that use the divided structure. In all cases, the division decision-maker will have to take into account the computational load of further division both to the division and downstream processes.

The advantage of using a complicated surface for division is that, at each division stage, a greater reduction in the size of the model is possible. This should result in more efficient processing of these sub-models by the downstream process. However, this advantage may be outweighed by the additional computational load required to generate the divided structure, and to process it.

1.5 Classification of polynomials against planes and other surfaces

The generation of a divided structure requires the classification of model primitives against sub-volumes. A primitive may either contribute entirely 'air' or solid to a sub-volume or may lie completely or partly inside the sub-volume. In the case of cuboid sub-volumes and planar primitives this classification may be achieved simply by testing two opposite corners of the sub-volume with respect to the primitive. If non-planar primitives are used then the classification is more difficult.

One method of classifying curved primitives against sub-volumes is to calculate the intersection curves of the primitive with the faces of the sub-volume. If intersections are found then the primitive obviously intersects the sub-volume. If there are no intersections then the primitive lies either entirely inside or outside the sub-volume. Which of these cases is true may be discovered by testing a single point on the surface of the primitive with respect to the sub-volume. The calculation of intersection curves is discussed later in the paper. It is worth observing here that the simpler the form of the sub-space faces, the less the computational load required to perform the tests.

It should be noted that for pruning purposes, exact classification is not required. The approach used in the latest modeller written at Bath [29] is to apply a 'conservative' classification that guarantees not to classify primitives as non-intersecting if they actually do intersect the sub-volume, but may (falsely) classify them as intersecting even if they do not. The affect of these conservative techniques is to give a lower efficiency in downstream applications of the divided structure than that which would be obtained by using an exact classification technique, if such an exact technique were possible.

1.6   Ray-casting: root finding

The basic requirement of the ray-caster is to determine if the ray intersects the model, and if so to generate one or more of the intersection points. The ray-caster can use the divided structure by operating in two stages: in the first stage the sub-volume tree is descended and when a sub-volume containing part of the model is reached, the intersections of the ray with the sub-model primitives are generated. Thus it is necessary to be able to calculate the intersections of vectors with those surfaces used in the model primitives, and also those used for the division process. If the primitive is represented by a polynomial inequality the parametric ray equation may be substituted into the polynomial half-space expression to generate a polynomial equation in the ray parameter; the roots of this equation give the intersections of the ray and the half-space.

If this polynomial is of order less than or equal to four, direct solution is possible, though not necessarily efficient or accurate (see the discussion of the quadratic equation in [6]). If it is of higher order, then an iterative technique must be used. There is a range of well-known numerical techniques available for calculating the real roots of a polynomial with real coefficients. However, it is worth noting that the requirement for ray-casting is to find roots only that lie within a given region. This may be a region defined by the extent of the object-space, or, in the case of a spatially-divided structure, may be as small as a single sub-volume. (It may be even more constrained by knowledge of the roots of other primitives that have already been found.) Most general-purpose root finding techniques are not able to take advantage of this. One of the great problems of numerical root-finding is distinguishing a double root, or 'nearly double' roots, from complex roots with very small imaginary parts. But this is precisely the situation encountered when a ray-caster encounters the horizon of, say, a sphere. The instability of numerical root-finding was pointed out by Wilkinson [22], and the range of symbolic techniques available for avoiding this instability is surveyed by Davenport [5].

1.7 Line and section drawing

As well as the need to produce shaded pictures, generally performed by ray-casting, there is also a requirement for 'line' drawings and the ability to take sections through models. One reason for this is that engineers have traditionally used engineering drawings to communicate ideas, and are likely to do so for many years to come. Also, line drawings of models may be used as a quick methd of feedback to the model creator [28].

In order to produce line drawings, the edges of the set-theoretic model have to be generated. Thus the modelling system has to be capable of calculating the intersection curves of any pair of model primitives. Also, the 'horizon lines' of curved primitives have to be found. The techniques of generating these discriminants and resultants are discussed in the following section. Line drawings of curved surfaces may be difficult to interpret, as only the edges of the surfaces are shown. Some modelling systems have avoided this problem by using surfaces that are represented by parametric patches, and faceted by drawing isoparametric lines. If the line drawings are used solely for visual interpretation, it may be sufficient to generate only approximations to the actual edges.

1.8 Useful symbolic algebra techniques

Testing for the intersection of a line with a surface

Often in solid modelling we need to know whether a given line passes through a surface, and if so, at which points. In fact a fast method for doing this is prerequisite for any ray-casting query. A three-dimensional line can be written

$$x = a + td$$
$$y = b + te$$
$$z = c + tf,$$

where t is the variable parameter of the line and $a, \ldots, f$ are constants. Substitution of these values into the equation of a surface $P(x, y, z) = 0$

yields a univariate polynomial in t. The points on the line where it meets the surface are then given by the roots of the polynomial in t. This process is normally performed manually, by the program author, but we intend to perform this (and many other) operations automatically with our algebra system. Root-finding is an expensive process and much time can often be saved by checking beforehand that there are indeed roots to be found. We can find the number of roots to a polynomial with a simple procedure involving a Sturm sequence which is described below. In the case of ray-casting this process is just what is needed: a fast way of rejecting surfaces which will not intersect the ray. If we cannot reject the surface on these grounds we can do a binary chop, using the Sturm sequence, to find exactly where the roots are.

Finding the number of real roots of a univariate polynomial: Sturm sequences

Given a univariate polynomial $P(x)$, the Sturm sequence procedure returns the number of real roots that exist between two limits, say $[a, b]$. We set $Q(x) = P'(x)$ then, with an initial list containing the two polynomials $P(x)$ and $Q(x)$, we repeatedly add to the list the negative of the remainder obtained when dividing the last polynomial but one by the last polynomial. The result is a list of polynomials of decreasing degree. We now substitute a for x in each polynomial in this list and calculate $S(a)$: the number of sign changes in the resulting list. Having done the same for b, we can find the number of real roots of $P(x)$ between a and b from $S(a) - S(b)$.

Eliminating a variable from two equations: resultants

A resultant is the single polynomial which results from the elimination of a variable from two polynomials which contain it. The result of eliminating the variable z from $P(x, y, z)$ and $Q(x, y, z)$ is written:

$$\text{res}(P(x, y, z), Q(x, y, z), z).$$

The resultant is a polynomial in x which is zero if and only if the two polynomials share a common zero for these values of x and y. The algorithm for finding a resultant is described by Davenport [5].

Testing for intersection between two surfaces

In many algorithms used in solid modelling work one wishes to be able to answer the question 'Does surface $P(x, y, z) = 0$ intersect with surface $Q(x, y, z) = 0$?' In particular, as has been mentioned, all the modelling systems written at Bath divide the model into simpler sub-models before attempting to answer the queries described above. Such a process needs to know whether a given surface lies wholly on one side of a plane or not.

The procedure for doing this runs as follows. First we take the z resultant of $P(x, y, z)$ and $Q(x, y, z)$. This eliminates the variable z from the two equations yielding a bivariate polynomial, say $R(x, y)$. We now need a procedure which will tell us if there are zeros of $R(x, y)$ which are real in both x and y. If there is none then the surfaces P and Q do not intersect. If there are, we have to determine whether the corresponding z values are real or not. For example, consider the intersection of a sphere

$$S(x, y, z) : x^2 + y^2 + z^2 - r^2 = 0$$

with a plane

$$P(x, y, z) : z - d = 0.$$

The resultant $\mathrm{res}(P(x, y, z), S(x, y, z), z)$ is:

$$R(x, y) = d^2 - r^2 + x^2 + y^2 = 0$$

which clearly has no real roots if $|d| > |r|$.

If this resultant were a univariate polynomial $R(x)$ rather than the bivariate polynomial $R(x, y)$ then the method of Sturm sequences described above would give a way of finding if roots existed. We have produced a fallback procedure for doing the required test on a bivariate polynomial. It seems, however, as if a two-dimensional generalization of the Sturm sequence method would be much more efficient, and this is therefore one of the subjects currently under research.

The generic ray

If ray-casting is to be used for picture production (one of its commonest applications) the time taken to generate a picture will be proportional to the area of the graphics screen on which the picture is to be displayed: that is, the number of pixels. Conventional methods of ray-casting do little more than simulate the path of light rays through each pixel on the screen and, due to the sheer number of pixels, turn out to be extremely expensive. One of the most appealing methods made possible by an algebraic approach is that of firing a generic ray. The generic ray is a single ray whose direction and starting point contain the screen coordinates, which are treated as variables in all the subsequent calculations. We can use the generic ray to map the three-dimensional set-theoretic model to a two-dimensional model in the screen coordinates. The two-dimensional model cuts the screen up into regions which are bounded by polynomials. The colour of each pixel within such a region is given by a single function and can therefore be evaluated very quickly.

The intersection of two surfaces $P(x, y, z) = 0$ and $Q(x, y, z) = 0$ in the object volume may yield a boundary on the screen. Each such boundary can be defined as polynomial equality in the screen coordinates:

$$f(s_x, s_y) = 0.$$

To find these polynomials we need to project the curve marking the intersection of $P(x, y, z)$ and $Q(x, y, z)$ on to the screen.

As an example of how this is done consider the parallel projection of the curve onto a screen aligned with the z axis. It turns out that, in this special case, $f(s_x.s_y) = \mathrm{res}(P(x, y, z), Q(x, y, z), z)$.

A general method for a perspective mapping to an arbitrarily positioned screen involves some more variables but reduces to a single resultant between two polynomials whose orders are the same as those of $P(x, y, z)$ and $Q(x, y, z)$. Such a mapping, which does all the intersection testing, is indeed likely to be much more expensive than the task of firing a single ray but needs to be done only once for the entire screen (or each subdivision of the screen), rather than once for each pixel.

1.9   Implications of algebra

The pros

Why propose using algebra in solid modelling systems when numerical techniques have had considerable success to date? The truth is that what virtues there were in numerical techniques have been diminished by the scale and complexity of problems currently being tackled in solid modelling.

The first big advantage of algebra over numbers is generality. The algebraic answer is an expression containing variables for which values can be substituted to find out about specific cases, but, more importantly, the algebraic expression allows us to ask more general questions of the model; it gives us a richer domain of discourse. For example, given two algebraic expressions defining two surfaces we can ask whether they intersect anywhere. Contrast this with the numerical technique which is only able to provide information about specific points—the ones at which the surfaces were evaluated.

The second advantage of using algebra is a corollary of the first;. we refer to accuracy. An algebraic solution is always exact; what errors that do arise come from the substitution of numerical approximations into the expression. Numerical accuracy is not a serious limitation when dealing with lots of plane surfaces (i.e. using planar half-spaces), but with the advent of blends to generate fillets many high-degree polynomials are generated which often induce instability in the numerical methods.

The cons

A disadvantage of algebra is a phenomenon probably unknown outside the algebra community, namely **intermediate expression swell**. This name is quite descriptive; it refers to the magnitude of the intermediate results that are generated on the way to an answer. It is an unfortunate consequence of the schemes by which most algebra systems operate that this is unavoidable. It means that a large amount of memory is used and then discarded, since often by the time that the final result is produced there have been many cancellations. It is for that reason we are concerned about the design of the storage management system, which is discussed in some detail in the next section of this paper. The growth happens because algebraic algorithm are relatively sensitive to their input, whilst numerical algorithms have much the same complexity regardless of input. In statistical terms the average behaviour (in terms of time and memory) of a numerical process has a low standard deviation; an algebraic process does not behave in the same way, and this is reflected in the number and size of the expressions generated in the course of the computation. This problem is not unassailable (see the section on additive complexity for instance) and can be tackled by clever representations and by the use of polyalgorithms.

The other detraction from algebra is the difficulty of writing algebraic programs. It demands three properties of the system builder:

- A high mathematical capability.
- An understanding of several esoteric issues in system design.
- An appreciation of algorithmic complexity and ingenuity in programming to circumvent it.

Our approach is to build a geometric algebra package which means that the majority of that complexity will be hidden behind a structured interface. All that will appear above that surface is a suite of routines for manipulating algebraic expressions. Further capabilities may be added to the library as the user becomes more experienced. We suspect

that the difficulties we have outlined are what might have discouraged
people from pursuing this approach in the past.

Sparsity versus density

How should the polynomials the system is manipulating be stored
internally? Is it necessary to store information explicitly about each
possible coefficient or is a more parsimonious representation in which
omitted terms are assumed to be zero more appropriate? The former
scheme is called **dense** and the latter **sparse**; the choice of representation
should be based on the kind of problems it is envisaged the system will
be asked to solve. In a general algebra system such as Macsyma [15]
or Reduce [7] there is a fairly clear-cut case for using a sparse
representation because, although it takes more memory to store each
element of the polynomial, the sparsity of most problems more than
outweighs that cost.

A related question on the matter of representation is whether it will
be necessary to manipulate multivariate polynomials or whether
univariate polynomials are sufficient. In the latter case it is possible to
build very highly-tuned systems with remarkable performance.
However, this is a case of 1 or n, since the move to bivariate and
trivariate leads to such messy code that it is better to use a general
representation. Clearly, in the domain of solid geometry one needs at
least three variables—but let us not prejudge the issue.

Some algebra systems provide as many polynomial representations
as the programmer wishes to design and support (e.g. Scratchpad [12],
Views [11]), but using them consistently would be next to impossible
without a sophisticated polymorphic typing scheme. Such a solution is
therefore way beyond the scope of this project where we are concerned
primarily with solid modelling system design and implementation and
only secondarily with the design and implementation of algebra systems.

- Dense representations: the most obvious dense representation is
  that of a vector of coefficients in which the index corresponds to the
  degree. This works quite well for univariate polynomials but the
  storage required rises dramatically with the number of variables (it
  is $O(n^m)$ where m is the number of variables and n is the maximum
  degree). The algorithms for the dense representation are relatively
  straightforward to write; in particular there is no need to worry
  about keeping track of the degrees of the terms being operated upon
  since the fact that they are compatible is guaranteed by position.

- Sparse recursive: as soon as one is concerned with sparse
  representations, vectors and arrays lose their appeal and a more

dynamic data structure is needed: the linked list. A very popular representation is the recursive structure (used widely in Macsyma and Reduce) where each element of the polynomial contains the variable, the exponent and the coefficient. The recursive nature comes from the fact that the coefficient need not be an integer (or whatever), but rather it can be a polynomial in a second variable, whose coefficients may be polynomials in a third variable, and so on. The system has to be very careful about maintaining such a structure, for instance a polynomial in n variables may be represented $n!$ ways which would make comparison very expensive. The solution is to require some ordering of the variables (e.g. lexicographic) so that $x + y$ will have an internal representation isomorphic to any other instance of $x + y$.

- Sparse distributed: an alternative sparse representation is the distributed form of an expression. Indeed for some algorithms the representation is critical, for example, Gröbner bases are very hard unless the polynomials are maintained in the distributed form. Informally one can think of the distributed form as the result of having multiplied all the brackets out of the recursive form. That is, rather than the coefficient of a term being an element of the ring over which the polynomial is constructed, the coefficient is only an element of the base ring (e.g. integers), and all the variables (regardless of exponent) with that coefficient are collected together. A concrete example is $x(1 + y)$; in the recursive form this is a polynomial in y with the coefficient x; in the distributed form it is a polynomial with two terms x and xy, each with coefficient 1 and all the exponents are 1. As mentioned above, some algorithms are sensitive to the polynomial representation, and therefore it is often necessary to provide more than one form in an algebra system, which then also involves the need for routines to convert from one form to another.

- Sparse tables: a representation developed for one of the older algebra systems, called ALTRAN [3,4,9], is also worth mentioning because it does not need any pointers. All the information is immediate. This restriction was placed on the designers by the fact that they were writing in FORTRAN. In view of the discussion of memory management and the difficulties of scanning to identify unused expressions, such a pointerless representation does have its attractions. Each polynomial is made of three parts: a vector of variables, a vector of exponents, and a two-dimensional table of coefficients. The vector of variables acts as the dope vector in one dimension of the coefficient table and the exponent vector in the other dimension.

- Additive complexity: the additive complexity of an expression is simply the number of plus (minus) signs that occur in it. The additive complexity of an expression has implications for how hard it is going to be to make a substitution into the expression and provides a rough measure of the cost. We mention additive complexity here not because there is a complete solution, indeed it is an area of active research, but rather to highlight a problem. For example, $x^{10} + x^5 + 1$ is a compact expression and is as compact as it can be, but now substitute $x + 1$ for x and the resulting polynomial in its expanded form is large. On the other hand $(x + 1)^{10} + (x + 1)^5 + 1$ is a minimal form if one uses additive complexity as the acceptance criterion.

- Straight-line programs [8] are one angle of attack on the additive complexity issue. In addition, straight-line programs give complete control of the structure of the algebraic expression to the user, although the system is free to manipulate (and optimize) the representation when computing with it. The output from the system will be another straight-line program. The key thing to notice is that the representation is not canonical—there is no structure in the internal representation—rather the intention is to mirror the structure of the external representation. For example, it is easier for the user to recognize the form $(x + 1)^9$ than

$$x^9 + 9x^8 + 36x^7 + 84x^6 + 126x^5 + 126x^4 + 84x^3 + 36x^2 + 9x + 1,$$

particularly if it has been multiplied by another expression. If we turn again to the example of $x^{10} + x^5 + 1$ and the subsequent replacement of x by $(x + 1)$, using the straight-line approach the expression can be written as

$$b := (x + 1)$$

$$a := b^2 + b + 1.$$

Notice that we also capture the use of only one copy of $(x + 1)$ in this specification, indeed if one were to write

$$a := (x + 1)^2 + (x + 1) + 1$$

the system should optimize it to the preceding form. This work is quite radical but very promising. Most of the basic algebraic algorithms for this representation have been worked out, but there remains a lot more research to be done.

1.10 The storage management problem

When we came to investigate the question of a storage management strategy for the geometric algebra system, we were not surprised to discover that there was no one obvious solution, nor indeed any solution at all. Let us begin by reviewing the requirements for the system's storage manager, then describe various 'standard' strategies, and then our own solution.

Requirements

- Efficiency: we assume that the system will require a large amount of storage, both the amount in use at any one time and also the turn-over rate of storage, as polynomials are created, substituted, roots found etc. As an example of this, we note that Reduce on the HLH Orion (a machine comparable with a VAX 11/780) can turn over a megabyte of storage in about 20 seconds CPU, and we would suppose a SUN 3 (the project development vehicle) to be somewhat faster. We can expect the requirement to be for relatively small amounts of memory (e.g. individual polynomials or big numbers), and hence that there will be a large number of allocations (and de-allocations, whether these are explicit or implicit).

- Portability: we intend to develop the software in C, and to compile it using a 'standard' compiler. If the system is the success we confidently predict, there will be a substantial requirement for portability. This certainly implies that large quantities of machine-code are not acceptable. More subtly, it means that we cannot rely on the features of a particular C compiler or run-time system.

- Robustness: it is possible to do all the storage management 'by hand', with the programmer taking great care to ensure that each object is freed at precisely the right moment. Such a scheme is likely to meet the criterion of efficiency, and probably that of portability. On the other hand, such programs are extremely difficult to write and debug, since an error signalled by any given function may well have arisen because a quite different piece of code freed a data item that to which the given function still had a pointer. Failure to free all the storage can lead to obscure 'leakages' of memory, and ultimately to the necessity of poring over a complete memory dump. In a multi-author project that is building a library for general use, the difficulty of enforcing the necessary conventions seem so great that we were very reluctant to follow this route.

Standard solutions

Before outlining the various possible techniques of storage management, it is useful to clarify some terminology. The **stack** is that area of memory used by C for 'auto' storage allocation, and also for compiler-generated temporary variables, return addresses and so on. **Global variables** are those objects of storage class 'static': in general we will ignore the problems they pose, since they are logically equivalent to a bottom layer of the stack (although there are substantial administrative problems associated with knowing where all the global variables are). The **heap** is that area of memory out of which the storage allocation system allocates new objects, and where old objects reside until their memory is recycled. **Fragmentation** (often called external fragmentation) is the problem of not having a contiguous chunk of available memory large enough to satisfy a particular request, even though the total amount of free memory is sufficient.

There are various requirements that we can impose on the stack or the heap for our storage management system to work. Let us first consider the stack (and global variables). We will say that the stack is **strongly scannable** if we can detect precisely those objects on that stack that are pointers to objects in the heap. We use the term **weakly scannable** to mean that we can find all the pointers, but that we may also get some objects that are not actually pointers, e.g. floating-point numbers or large integers that just happen to be valid as addresses in the heap, even though that is not how they were created (this is related to the idea of conservative scan described by Swineheart et al. [18]). For the heap, we will use the same terms of particular objects in the heap, saying that they are strongly scannable if we can detect precisely the components that are pointers into the heap, and weakly scannable if we might get some objects that look like pointers but are not meant to be pointers. We will say that the heap is **sweepable** if it is possible to start at the bottom of the heap, and identify each object in the heap in turn.

We can now consider the various storage management techniques available, with their requirements, advantages and disadvantages.

- Reference counts: with each object, we associate a count of the number of objects that point to it, and free the object when this count reaches zero. This requires that heap objects be strongly scannable, since when an object is freed, we have to decrease the reference counts of all the objects to which it points (possibly freeing them, and so on). Weak scannability is not sufficient, since we might decrease the reference count of some object by mistake, or even tamper with something that was not a reference count at all. Fragmentation of the list of free objects is somewhat of a problem, but we can hope that an efficient recombination method will help here.

- Garbage collection with a free list: this technique requires that the stack and heap be weakly scannable, so that we can determine which objects are in use, and which are not. Strong scannability is not required, though it does make life easier. The heap also has to be sweepable, since, after the 'mark' phase of descending all objects whose address we find on the stack, we then sweep linearly up the stack, looking for unreferenced objects and placing them on the free list. Again, fragmentation is a slight worry.

- Compacting garbage collection: this is similar to the previous method, but the sweeping phase actually compacts all the in-use objects to one end of the heap. This is then followed by a relocation phase of adjusting all the pointers that point to objects that have been moved. Hence we need strong scannability of the stack and of heap objects, since it would be embarrassing to relocate something that was not actually a pointer.

- Copying garbage collection: this strategy employs two heaps, of which one is 'in use' at any time. When this heap is full, the stack is descended, and all accessible objects are moved to the other heap, and pointers to them are updated during the move. When this process is over, only part of the new heap should be used, and we can continue running from this heap. This process requires strong scannability of the stack and the heap objects, but does not require that the heap be sweepable.

Our solution

Having looked at the various strategies available, let us consider how we might implement them in C. Weak scannability of the stack is not an insuperable problem, since it is possible to take addresses of stack objects and then treat the stack as a vector. Strong scannability is much harder. LISP systems in general achieve this via type-coding of all data items, but this assumes that the compiler will not accidentally create objects with the same code. This is a reasonable assumption to make if one has complete control over the compiler, but we do not. With some care, weak and strong scannability of the heap objects can be assured, and in fact placing a uniform header at the start of each heap object will also make the heap sweepable.

Hence the usable strategies are reference counts and free-list garbage collection, unless drastic steps are taken to make the stack strongly scannable. Free-list garbage collection will suffer a performance penalty because of the number of false pointers on the stack. Swineheart et al. [18] report that this is not much of a problem in CEDAR, but their system and applications are sufficiently different from ours that a direct inference is dangerous. Furthermore, they have their own compiler, and have no real experience of moving to different architectures.

Reference counting is also a possible option, apart from the amount of work that the programmer has to go to in maintaining the reference counts. There is also the worry that both these systems might suffer from fragmentation, though this is probably not a very serious worry.

All these circumstances lead us to deduce that a definite decision so early in the life of the project was probably a mistake, and that we wanted to keep our options open. We therefore decided, as is often the case in computing, to postpone the decision by writing code portable between the different strategies, which can actually be done remarkably simply with the aid of a preprocessor. This preprocessor accepts a language (baptized P) which differs from C in one respect only, that there is a keyword heap. Hence the declaration

```
heap struct polynomial *fred
```

indicates that 'fred' is a pointer to a polynomial structure which is held on the heap. Depending on the strategy chosen, this information can either be used for adjusting reference counts or for maintaining a private copy of the stack which contains only the pointers into the heap, thus ensuring strong scannability.

1.11 Conclusions

This paper has outlined the scope of a new research project, whose necessity arises both from the computational imperative of applying algebra to CAD and the difficulties of applying existing systems to this problem. The authors are currently working on the detailed specification of the memory management system, and the model domain. During the life of the project the authors expect to investigate many of the topics discussed in the paper and to derive and implement techniques for solving them based on computer algebra.

Acknowledgements

The project for which this is an initial paper is supported by the SERC ACME Directorate. The work on blended models was also supported by a previous SERC grant. The authors would like to extend their thanks to the SERC for their support. They would also like to thank John Woodwark, who instigated the research on polynomially-based modellers at Bath, and who did a great deal of the initial work on this subject. Finally, they would like to thank Dayong Zhang for the work that he has done on the precursors of this project, particularly in the field of gap blends, and also for creating the models depicted in Figures 1.1, 1.2 and 1.3.

Appendix: Liming-Woodwark-Zhang polynomials

The problem of representing fillets has already been mentioned. Fillet or corner blend surfaces exist on many engineering components. They distinguish themselves in that they link other surfaces, which will be called blended surfaces hereafter, with only one blend for their cross-section shape; in other words, since the blended surfaces lie on the same side of the blend surface, they can in principle be dealt with by techniques such as those which try to reproduce the effect of rolling a ball along the joins between the blended surfaces. (This particular technique is hard, in practice, to implement.) A comprehensive review of this type of blend and the various means used to represent it can be found in Woodwark [26]. Chamfers, fillets and blends between intersected cylinders, the blend at an edge formed by two surfaces or at a corner of three or more surfaces, are all of this type.

The authors and their colleagues (notably John Woodwark) have developed a blend based on and developed from the definition of Liming quadrics for solving this problem. In aircraft fuselage design, Liming [13] used straight lines to specify quadratic curves for describing fuselage cross-sections. For instance, a quadratic curve Q (Fig. 1.4) can be defined as

$$Q : (1 - \lambda)AB - \lambda C^2 = 0,$$

where A, B and C are the straight lines

$$A = A_1 x + A_2 y + A_3$$
$$B = B_1 x + B_2 y + B_3$$
$$C = C_1 x + C_2 y + C_3$$

and λ is a constant in [0,1]. The quadratic curve Q links lines A and B together smoothly, with slope continuity at the intersections between A, B and C. Part of Q (the shaded part in Fig. 1.4, for instance) can be used as a blend curve between A and B. If A and B are treated as half-spaces, the blend can also be considered to be a half-space, and all three can be used to define a solid in a set-theoretic solid modeller. By intersecting Q with C, the blend can be truncated at exactly the places

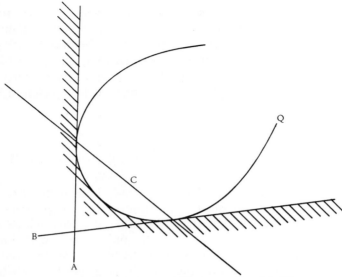

Fig. 1.4 Using Liming quadratics to blend half-spaces.

where it ceases to be useful. The set-theoretic expression describing the shaded area in Fig. 1.4 is:

$$A \cup B \cup (((1 - \lambda)AB + \lambda C^2) \cap C).$$

The range of the blend can easily be controlled by the position of line C, which is called the **ranging surface**, and the extent of the blend can be controlled by the value of λ. When λ changes from 0 to 1, the blend changes from A and B to C.

This method can be extended into three-dimensional space and to specify quadric surfaces by replacing the straight lines with planes. Furthermore, by replacing the planes with curved surfaces, and changing the number of blended surfaces and ranging surfaces, blends can be specified that link more complicated surfaces together smoothly. When surfaces defined in this way are used for purposes other than blends, this method also provides a general way to define high-order implicit polynomials. By using surfaces defined by one equation to define another, much more complicated surfaces can be represented and blends can be constructed on previous blends. Different versions and generalized forms of the Liming techniques have been used in different solid modelling systems [14,29].

Detailed descriptions and discussions about corner blends are given in the indicated references and one of the authors' student's Ph.D. thesis [31].

Corner blend surfaces specified by the above techniques have proven effective and convenient to use in solid modelling. However, there are

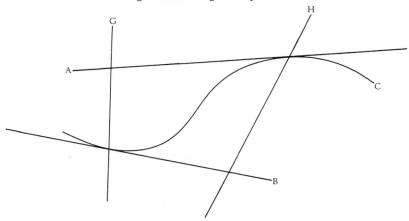

Fig. 1.5 The Zhang polynomial for a gap-blend.

other types of blend surfaces (which also appear on many engineering components and other objects) that are difficult to represent by corner blends; for instance, the blend between the two concentric cylinders of different diameter at the neck of a wine bottle. In such a situation, the blended surfaces generally lie at different sides of the blend and cannot be blended together by rolling-ball and similar techniques.

Since the blend has to be continuous with both of the blended surfaces, the corner blend is not suitable for this and a different method has to be used. These blends contain inflections and may be called, intuitively, gap blends. Armed with a blend of this type, a modelling system can represent many different pipes and ducts with smooth transitions from one cross section to another, and also sheet metal work with twists and other complicated deformations [33].

To begin with, consider blending two lines in two dimensions. Suppose that lines A and B are to be blended within the area defined by G and H, which are another two lines, as shown in Fig. 1.5. The blend is required to link A and B together smoothly with tangent points on A at the intersection of A and G and on B at the intersection of B and H. The equation

$$C : (1 - \lambda)AH^2 + \lambda BG^2 = 0$$

defines a curve C, which meets the requirements. λ is a constant within the range [0,1], as before.

This method, invented by Dayong Zhang, can also be extended into three-dimensional space to blend planes. In three dimensions, more complicated relationships between planes can be established, and, not only can planes be blended in the same way as blending lines in two dimensions, they can also be twisted and curled as shown in Fig. 1.3. (For details of the rendering algorithms used to produce the pictures in this paper, and also of algorithms that the authors use to generate cutting tool paths from the solid models, the reader is referred to previous papers [23,32].) As before, curved surfaces can be blended just as easily by substituting their equations for those of the planes.

References and bibliography

1. S.K. Abdali, G.W. Cherry, and N. Soiffer, 'An object-oriented approach to algebra system design', Proceedings of SYMSAC '87, ACM, New York (24-30), 1986.

2. A. Bowyer, 'SID: A language for describing solid objects. User's manual', Bath University, 1985.

3. W.S. Brown, 'On computing with rational expressions', Proceedings EUROSAM 74 (26-34), 1974.

4. W.S. Brown, 'ALTRAN user's manual', Bell Labs, 1977.

5. J.H. Davenport, 'Computer algebra for cylindrical algebraic decomposition', KTH Stockholm, NADA Report TRITA-NA-8511, September 1985.

6. J.H. Davenport, 'Survey of symbolic applications for numeric computation', Project DIAMOND Paper 03/2-7/B01.p.

7. J.P. Fitch, 'Solving algebraic problems with REDUCE', *Journal of Symbolic Computation* 1 (211-227), 1985.

8. T.S. Freeman, G.M. Imiriziani and E. Kaltofen, 'A system for manipulating polynomials given by straight-line programs', Proceedings of ACM SYMSAC '86, New York (169-175), 1986.

9. A.D. Hall, 'Factored rational expressions in ALTRAN', Proceedings of EUROSAM 74, (35-45), 1974.

10. C. Hoffmann and J. Hopcroft, 'Automatic surface generation in computer aided design', Cornell University Computer Science Department Technical Report 85-661, 1985.

11. J.E. Hopcroft, 'The impact of robotics on computer science', *Communications of the ACM* 29 (486-498), 1986.

12. R.D. Jenks, 'A primer: 11 keys to new SCRATCHPAD', Proceedings of EUROSAM 84, (LNCS #174, Springer Verlag) (123-147), 1984.

13. R.A. Liming, *Practical Analytical Geometry with Applications to Aircraft*, Macmillan, London, 1944.

14. A.E. Middleditch and K.H. Sears, 'Blend surfaces for set-theoretic volume modelling systems', Proceedings of ACM SIGGRAPH 85, San Francisco (*ACM Computer Graphics* 19) (161-170), 1985.

15. R. Pavelle and P.S. Wang, 'Macsyma from F to G', *Journal of Symbolic Computation* 1 (69-100), 1985.

16. M.B. Phillips and G.M. Odell, 'An algorithm for locating and displaying the intersection of two arbitrary surfaces', *IEEE Computer Graphics and Applications* 4,9 (48-58), September 1984.

17. T.W. Sederberg and D.C. Anderson, 'Implicit representation of parametric curves and surfaces', *Computer Vision, Graphics, and Image Processing* **28**,1 (72-84), 1984.

18. D.C. Swineheart, P.T. Zellwger, R.J. Beach and R.B. Hagmann, 'A structural view of the Cedar programming environment', *ACM TOPLAS* **8** (419-490), 1986.

19. A.F. Wallis and J.R. Woodwark, 'Interrogating solid models', Proceedings of the CAD 84 Conference, Brighton, Butterworths, London (236-243), 1984.

20. A.F. Wallis and J.R. Woodwark, 'Creating large solid models for NC toolpath verification', Proceedings of the CAD 84 Conference, Brighton, Butterworths, London (455-460), 1984.

21. D.B. Welbourn, 'Full three-dimensional CADCAM', *Computer-Aided Engineering Journal* **1** (Part 1: 54-60, Part 2: 189-192), 1984.

22. J.H. Wilkinson, 'The evaluation of the zeros of ill-conditioned polynomials', *Numerical Mathematics* **1** (150-180), 1959.

23. J.R. Woodwark and A. Bowyer, 'Better and faster pictures from solid models', *Computer-Aided Engineering Journal* **3** (17-24), 1986.

24. J.R. Woodwark and K.M. Quinlan, 'Reducing the effect of complexity on volume model evaluation', *Computer-Aided Design* **14** (89-92) 1982.

25. J.R. Woodwark and K.M. Quinlan, 'The derivation of graphics from volume models by recursive subdivision of the object space', Proceedings of the CG 80 Conference, Brighton, England (335-343), 1980.

26. J.R. Woodwark, 'Blends in Geometric Modelling', in *The Mathematics of Surfaces* (R.R. Martin, ed.) (Proceedings of the 2nd IMA Conference on the Mathematics of Surfaces, Cardiff, September 1986), OUP (255-297), 1987.

27. J.R. Woodwark, *Computing Shape*, Butterworths, London, 1986.

28. J.R. Woodwark, 'Generating wireframes from set-theoretic solid models by spatial division', *Computer-Aided Design* **18** (307), 1986.

29. J.R. Woodwark and A.F. Wallis, 'A solid modelling system based on implicit blends', **in preparation**.

30. J.R. Woodwark and A.F. Wallis 'Graphical input to a Boolean solid modeller', Proceedings of the CAD 82 Conference, Brighton, Butterworths (681-685), 1982.

31. D. Zhang, 'Solid modelling and NC machining of blend surfaces', PhD Thesis, Bath University, 1986.

32. D. Zhang and A. Bowyer, 'CSG set-theoretic solid modelling and NC machining of blend surfaces', Proceedings of the 2nd ACM Symposium on Computational Geometry, New York, June 1986.

33. D. Zhang and A. Bowyer, 'A new type of blend surface for solid modelling', **in preparation**.

Discussion

(This paper was presented by Milne.)

Middleditch: Could you expand on the reasons for not using Pascal?

Milne: Ordinary versions of Pascal are not suitable for writing big programs.

Bowyer: Pascal is fine if you want to write one piece of monolithic code, but makes it quite difficult to write modular programs. There are versions which allow you to write modules and link them together but you do get portability problems to a certain extent. I think C is a more portable language than Pascal; it has a slightly richer set of operations. It allows you to get your hands dirty in ways that Pascal does not, and sometimes that is necessary.

Davenport: As a rider to that, I actually wrote an algebra system in Pascal, and never wish to again!

Todd: You haven't yet written one in C. Perhaps when you have you will make a similar comment about that language!

Fisher: Why not buy a LISP machine?

Milne: Portability. There's no standard LISP.

Fisher: How about Common LISP?

Davenport: I wish we could write this system in LISP—it would be much easier. The facts of life are that most designers using CAD don't have LISP machines, nor machines supporting anything approximating Common LISP. If the world was all run in Common LISP, then it would be a natural choice. The world's not all run in Common LISP, I don't believe that that situation will change within the next five years.

Fisher: Would it be worth considering LISP for prototyping, and then re-programming later?

Davenport: We have prototype algebra systems already: Macsyma, for instance.

Geisow: Can I get away from languages and back to the main topic? You mentioned solving roots of bivariate equations, and then just gave a method for univarate ones.

Milne: What we've got at the moment is a fall-back procedure for doing bivariate root-finding, which has been thought of by James Davenport [34]. At the moment it looks as though, if we can generalize the Sturm sequence, we should be able to get a method that's much faster than this, but this will do the job, if we can't do that.

Forrest: Two things worry me. One is the use of ray-casting to produce your line drawings. Can't you find a better way than that, because the lines will be horrible. Secondly, aren't you severely restricting yourself when you limit yourself to integer coefficients? You must be missing a lot of interesting shapes.

Milne: To answer your first question, this isn't really ray-casting. Ray-casting involves choosing a point on the screen, and firing a ray through it. Our proposed technique is to treat a point on the screen as a variable, so it's not really ray-casting at all, it's mapping three-dimensional equations on to two-dimensional ones on the screen. There won't be any problem then in tracing those out—they won't look horrible.

Forrest: Can't you use a more direct way of finding silhouettes?

Milne: It is effectively finding a perspective silhouette, by saying all the rays are going to meet in the eye. With regard to your second question, we are using rational, rather than integer, coefficients.

Forrest: Even rational coefficients imply a limitation to the range of shapes that can be dealt with.

Davenport: But if I take a denominator of 2^{24} I've included all floating-point numbers.

Forrest: Still not good enough for some things.

Davenport: Well, take 2^{46} as the denominator and that gives all double precision numbers!

Wallis: I think a point that should be made is that you can draw these things with numerical techniques. One of our considerations is to get around numerical problems by using algebraic techniques rather than double or even quadruple precision arithmetic. It's not that you can't do these things numerically—a lot of people here have already done them—but we feel that the algebraic techniques are well worth looking at and have a number of advantages.

Jared: You're essentially providing a geometric service on the top of which you might construct a geometric modeller. Have you any feel whether you can do these geometric calculations sufficiently quickly for the modeller on top to run at a reasonable speed? Also, since you are using rationals, do you feel that the program will run out of even the enormous virtual memories on current workstations?

Davenport: To the first question—no. That's why it's research, not development. However, I believe that it will end up running faster, if only because, as technology advances, every time you double your screen resolution, anything that does four times the number of pixels inevitably runs four times as slowly, whereas the amount of algebra to be done is the same. But it's research; I don't guarantee that it will run faster; when the project is over I'll tell you.

In answer to your second question, no, I don't think so—there are good ways to economize in storing rational numbers: but again, it's a worry—I don't think it'll happen like that, but we'll see.

Bowyer: A point worth emphasizing, although it's been made already, is that if you generate an algebraic answer to a problem then, in a sense, it is an answer to an infinite number of problems. A numerical answer is an answer to that problem only. If you generate answers algebraically to the problem of generating an image, than those are very general answers, and you can apply them very speedily to the problem of producing a slightly different image.

Jared: I think the aim of doing it all by algebra is eminently praiseworthy, having had to depend on a numerical package, for a number of years. I'd like to have your system *now*, please!

Bowyer: One thing we should mention, is that we'd be very interested in people's opinions of what should go into this geometric alegbra system. We have our own ideas, of course, which are generated largely by the problems which we perceive in our own work, but we would be interested in other people's ideas on this subject.

Jared: You seem to be providing services for a CSG modeller. Would you also like to provide services for a boundary one?

Bowyer: If you'd like to make a list, we'll certainly consider it.

Additional References

34.　　J.H. Davenport, 'Computer algebra for cylindrical algebraic decomposition', University of Bath, Department of Computer Science Technical Report 88-10, 1988.

2 A high-level language environment for the definition and manipulation of geometric forms

H. CRAPO and J.-F. ROTGÉ

2.1 Computational tools for geometric research

Geometric research has long been impeded by the unavailability of adequate and efficient means for visual representation, and by the unavoidable gap which separates concrete geometric models from their logical and algebraic description. Recent advances in computer-aided design, together with progress in computer-aided geometric reasoning, promise a speedy improvement in the conditions under which geometric research is carried out.

Since computers have not been told that there is anything special about three dimensions, they are perfectly content to work on higher-dimensional problems, when programmed to use the usual techniques of vector representation and linear algebra. Higher-dimensional sub-spaces are easily represented in projective (Grassmann-Plucker) coordinates. Exterior algebra, suitably upgraded to the Rota-Doubilet-Stein double algebra of *join* and *meet* [10], enables one to draw out the consequences of geometric hypotheses for geometric figures of arbitrary dimension. But since computer *output* is typically no more than two-dimensional, consisting as it does of essentially one-dimensional strings of letters, and two-dimensional drawings or displays, some adequate way has to be found to represent and manipulate higher-dimensional structures in two-dimensional form.

2.2 A glimpse of descriptive geometry

Since the time of Gaspard Monge (1746-1818), geometers have successfully developed techniques to bridge the gap between two and

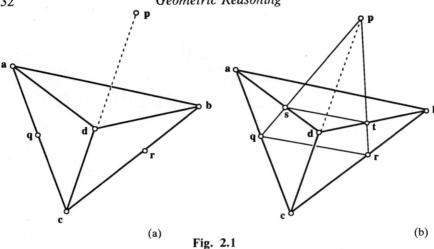

(a)   (b)
Fig. 2.1

three dimensions. The most substantial effort goes by the name of
descriptive geometry, a sort of applied Euclidean geometry. The basic
technique in descriptive geometry is to work on plane drawings as if
they were already three-dimensional. The first step is just a question
of correct labelling of a plane figure: the visible intersection of two lines
in the drawing plane is not taken to be a point unless the two lines are
known to be coplanar in the associated spatial realization. As a second
step, the descriptive geometry technique of creating views by the
rotation of flat polygonal faces down into the plane can be used to
obtain correct Euclidean dimensions for plane faces, so they can be cut
from cardboard, and real three-dimensional models can be built.

We illustrate these two principles. In Fig. 2.1 we have the problem,
how to find the intersections of the plane determined by points **p, q, r**
with the tetrahedron **abcd**. Since the lines **cp** and **bd** are coplanar, the
apparent intersection **t** of line **pr** and **bd** is a real point. Points **q, r, s**
and **t** determine the polygon of intersection. By the way, the lines **ab**,
st and **qr** are concurrent. Why? Inclusion of that tenth point (where the
three lines meet) completes what is known as a **Desargues
configuration**, the plane projection of the 10 pairwise intersections of five
general planes in space.

Now trim away that part of the tetrahedron which lies 'behind' the
cutting plane. We have a correct drawing (Fig. 2.2) of a triangular
prism. The fact that lines **qs, cd, rt** are concurrent follows from the
descriptive geometry construction, and is a necessary and sufficient
condition that the drawing of the prism be correct. A drawing in which
this condition fails will have no consistent three-dimensional
interpretation! The study of such projective conditions is the *sine qua
non* of scene analysis. It is also crucial to an understanding of the
infinitesimal mechanics of bar-and-joint structures. Theorems of

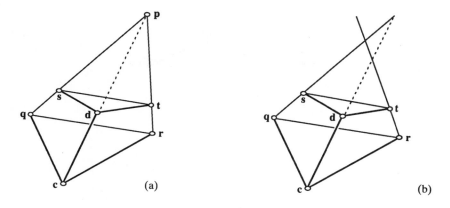

Fig. 2.2   Plane drawings with (a) and without (b) and three-dimensional realization.

Maxwell and Cremona relate three-dimensional realizations of polyhedral graphs to stresses in plane frameworks, and thus to unexpected degrees of freedom for infinitesimal motions. For the example in Fig. 2.2, the correct drawing gives a plane bar-and-joint structure with one degree of infinitesimal motion, moving one triangle relative to the other. The spatially incorrect drawing is a correct drawing of an infinitesimally rigid framework.

Descriptive geometry constructions of intersections are not always so straightforward. Fig. 2.3 illustrates the sort of problem. How can we construct the intersection figure of the plane determined by given points on three faceplanes of the corner of a tetrahedron (three planes **A**, **B** and **C** meeting at a point)? The construction starts with an arbitrary point **p** on one of the three lines (say **A** ∩ **B**), and continues by locating the point **q** where the line **ab** pierces the plane **C**. Starting from point **q**, a polygonal path through points **c**, **a** and **b**, is bound to close at a point on line **B** ∩ **C**, because successive edges are forced to be the intersections of the plane **abc** with faces **C**, **A** and **B** respectively.

In Fig. 2.4 we construct a view of the triangular prism. We select as base plane the plane of face **qrst**. Imagine that each of the four other faces are rotated down from their position in space, into the base plane, using the common edges as hinges. The path of vertices **c**, **d** will be seen in projection as straight lines perpendicular to the corresponding hinges. We are free to choose one of these images of **c** or **d** at will; the rest will then be determined. The essential fact is that a line like **pd** will be the same length in each of its rotated images. Thus length(**pd′**) = length(**pd″**) and the position of **d″** determines the position of **d′**. Furthermore, the collinearity of **p**, **d**, **c** will be preserved in the image. This determines the location **c′**, and similarly for **c″** . The point **d‴** is

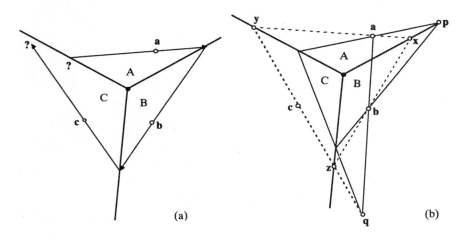

Fig. 2.3

located by the conditions that length(**sd′**) = length(**sd‴**) and length(**td″**) = length(**td‴**); a similar calculation locates **c‴**. The final view is such that, if the plane figure be folded along the hinge lines, the faces become the face planes of a true spatial polyhedron (that is, the points **c′, c″, c‴** and **d′, d″** and **d‴** coincide respectively at two points **c** and **d**).

There is no obstacle, at least in theory, to extending the methods of descriptive geometry to higher dimensions. Janos Baracs, a colleague of ours and founder of the Structural Topology Research Group in Montreal, undertook the extension of descriptive geometry to four and five dimensions, in order to understand the mechanics and statics of bar and joint structures in three dimensions. One example will suffice to show the sort of thinking which is involved. Imagine (Fig. 2.5) you are given four concurrent lines $L_1, \dots L_4$ in 3-space, and on each line L_i two points a_i, b_i, ($i = 1, \dots 4$). Construct two skew polygons (quadrilaterals) P_a and P_b, one through the points a_i, the other through the points b_i. It is a theorem of real projective geometry that the four points

$$p_i = (a_i \vee a_{i+1}) \wedge (b_i \vee b_{i+1}) \text{ for } i = 1, \dots , 4 \text{ (modulo 4)}$$

are coplanar. Baracs' proof proceeds via a construction in four-dimensional space, as follows. Let the four lines $L_1, \dots L_4$ extend in four independent directions from a point in 4-space **S**. Then the points in the polygons P_a, P_b generate two 3-spaces T_a, T_b. These 3-spaces T_a, T_b are distinct, and are contained in a common 4-space, so their intersection is two-dimensional. (This law of modularity of dimensions is the basic feature of projective geometry.) The two-dimensional space $T_a \wedge T_b$ contains all four points p_i: they are coplanar in 4-space. If this

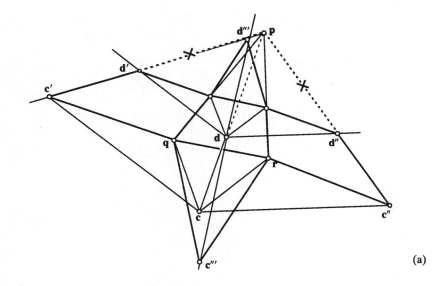

(a)

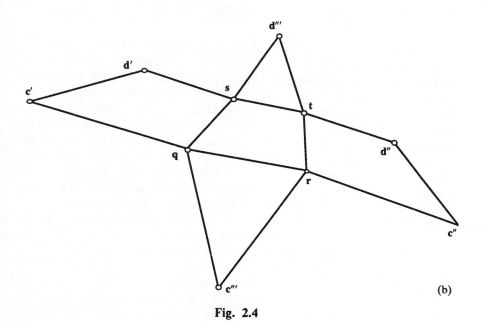

(b)

Fig. 2.4

entire figure is now projected into 3-space, the image of the space $T_a \wedge T_b$ will still have dimension at most equal to 2, so the four points P_i will still be coplanar, as required.

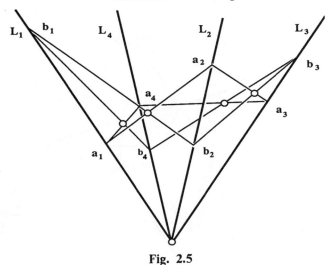

Fig. 2.5

2.3　The use of partially defined objects

The natural description of geometric structures, and thus the starting point for any comprehensive geometry software package, is in terms of **variable** points, and **combinatorial** statements of incidence and other projective properties. The corresponding calculations can only be accomplished in terms of **polynomials** in the coordinates of undetermined points. Symbolic computation methods are applicable to this problem, but will need to be extended to cover the gap mentioned in our introductory paragraph. Geometric statements, correctly translated into geometric language, do not necessarily describe *single* families of geometric models. There is a phenomenon of **branching** omnipresent in the adjoint situation linking geometric properties and geometric models. Algebraically, this is simply the observation that an radical ideal is an **intersection** of prime ideals.

A simple example will suffice to underline the gravity of this situation. Say we want a figure of five points **a**, **b**, **c**, **d** and **e** in the projective plane, such that **abc**, **ade** and **bce** form collinear triples. One way to form a maximally general model is to draw a figure with all five points collinear. This can be done with seven projective degrees of freedom. But another entirely different figure is possible. Let the points **c** and **d** coincide: then there is no longer any need for the combined point **c** to be on the line **ade**. This figure can also be drawn with 7 projective degrees of freedom. A more adequate idea of the richness of this branching process in more complicated figures can be gained even

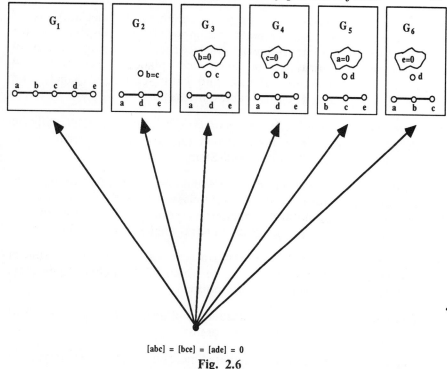

[abc] = [bce] = [ade] = 0

Fig. 2.6

here, if we admit that certain points may be zero (like points at a centre of projection, under projection). Thus, if **a** is zero, there is no longer any need for **d** to be on the line **bce**. Fig. 2.6 shows the six branches for minimal models of the description $|\textbf{abc}| = |\textbf{ade}| = |\textbf{bce}| = 0$, the vertical bars indicating determinants.

2.4 Outlines of a programming environment

The essential features of a programming environment for automated descriptive geometry would seem to include the following:

- Input devices flexible enough to generate exact data of incidence, approximate data of location, without impeding the free use of geometric imagination.

- A database permitting the stocking of a variety of partially defined geometric figures, either generic, or else with specific projective coordinates, whenever these become available.

- A rich vocabulary of elementary forms.

- Interactive definition of complex structures, by declarations of incidence or by other methods of composition starting from simple structures (including splines). Such structures should include all sorts of configurations of lines and planes, mechanical and architectural structures of bars and joints, of hinged panels, and tensegrity systems of elastic cables or sheets.

- Calculation of the effect of geometric operations, principally of projection and intersection, carried out in symbolic form in the double algebra of Doubilet-Rota-Stein.

- Automatic construction of generic models (often after taking into account the natural branching into classes of minimal models) and interactive computation of models determined by a series of free choices of heights, bar lengths, dihedral angles, and the like.

- Automatic calculation of descriptions of figures with special geometric properties, such as those which lift to higher dimensions, or which admit certain internal motions.

- Derivation of the logical consequences of geometric hypotheses, and automatic proof of geometric theorems.

- Legible representations of higher-dimensional geometric forms, via screens, plotters and laser printers.

2.5 Some initial decisions

The present project was recently undertaken in collaboration between the Institut National de Recherche en Informatique et en Automatique (INRIA) and the company SERBI, creator and distributor of the CAD software, Conception 3D. We have decided to take advantage of the capacity of the language LISP (Le-LISP, Common LISP, etc.) to handle partially-defined structures. The great advantage of this family of languages is that variables do not have to be fully evaluated until such evaluation is possible. The form that a value will produce does not even have to be specified in advance.

It is clear that the analysis of geometric properties and the automated proof of geometric theorems will also involve the treatment of polynomial rings and their ideals, an appeal to standard bases or cylindrical algebra decomposition, as practiced in the Macaulay software, or in the work of Wu [e.g. 16], Collins [e.g. 4], McCallum, Chou [e.g. 3], Buchberger [e.g. 2], Kutzler and Stifter [e.g. 11] and others, often linked to Scratchpad or other systems for symbolic computation. We do not as yet know whether the presently available software can handle the search for minimal prime ideals containing a given radical ideal, and thus whether it can produce generic geometric models of given hypotheses.

A systematic use of projective coordinates avoids a good portion of the work geometers are required to do in order to classify and isolate the many degenerate cases in which a generic geometric truth fails to hold. (For a good example of care taken in the presentation of hypotheses of non-degeneracy, see Chou [3].) Points, lines, sub-spaces of all dimensions, vectors, forces, moments, rigid motions, etc. all have compact representations in projective coordinates. Incidence of flats, and dependence of point sets, are easily expressed in terms of projective coordinates.

It is becoming apparent that the use of intermediate combinatorial structures can greatly simplify geometric programming. The chief use of such structures, such as matroids (combinatorial geometries) or their oriented counterparts, is to provide a preliminary classification of possible solutions to a geometric problem. Bokowski and Sturmfels [e.g. 12] have solved a number of long-outstanding problems concerning geometric realizability of combinatorially-defined polytopes. They have shown that the search for geometric realizations becomes feasible only when the search is restricted to a single orientation class. The preliminary listing of such classes is carried out using combinatorial algorithms and the theory of oriented matroids.

2.6 Applications

I have emphasized that the goal of the project is to create tools for geometric research. This is not at all to denigrate the use of such methods for practical pursuits such as computer-aided design, animation, three-dimensional modelling in (bio)chemistry, robotics, etc. It is simply to suggest that tools for geometric reasoning by computer, if they are adequate to the task of theorem-proving in higher-dimensional projective geometry, will also possess the flexibility to be of quite general use in a wide variety of practical settings. In conclusion, I would like to propose the a test of performance. Such software should be capable of describing, in geometric language, the projective conditions under which a bar-and-joint structure becomes unexpectedly flexible, or under which a two-dimensional picture is the correct projection of a polytope in higher dimensions.

References

The following references comprise a suggestive, but very incomplete bibliography.

1. D. Bayer and M. Stillman, 'Macaulay users manual', Draft 1.1 (D. Dayer, Columbia University, M. Stillman, Brandeis University), March 1986.

2. B. Buchberger, 'Gröbner bases: An algorithmic method in polynomial ideal theory', *Multidimensional Systems Theory* (N.K. Bose, ed), D. Reidel, Dordrecht (184-232), 1985.

3. S.-C. Chou, 'Proving geometry theorems using Wu's method: a collection of geometry theorems proved mechanically', **preprint**, Institute for Computing Science, University of Texas at Austin, 1986.

4. G.E. Collins, 'Quantifier elimination for real closed fields by cylindrical algebra decomposition', Proceedings of the 2nd GI Conference on Automata and Formal Languages, Lecture Notes in Computer Science 33, Springer (134-163), 1975.

5. H. Crapo, 'Structural rigidity', *Structural Topology* 1 (26-45) 1979.

6. H. Crapo, 'The combinatorial theory of structures: lectures on the application of combinatorial geometry in architecture and structural engineering', Proceedings of the Colloquium on Matroid Theory, North Holland, Amsterdam, 1984.

7. H. Crapo and J. Ryan, 'Spatial realizations of linear scenes', *Structural Topology* 13, 1986.

8. H. Crapo and W. Whiteley, 'The statics of frameworks and movements of panel structures, a geometric introduction', *Structural Topology* 6 (43-82), 1982.

9. H. Crapo and W. Whiteley, 'Plane stresses and projected polyhedra', *Structural Topology,* **to appear**.

10. P. Doubilet, G.-C. Rota and J. Stein, 'On the foundations of combinatorial theory IX: Combinatorial methods in invariant theory', *Studies in Applied Mathematics* 53 (185-216), 1974.

11. B. Kutzler and S. Stifter, 'New approaches to computerized proofs of geometry theorems', **presented at** a Conference on Computers and Mathematics, Stanford, 1986.

12. B. Sturmfels, 'On the decidability of Diophantine problems in combinatorial geometry', **preprint**, University of Washington, 1986.

13. K. Sugihara, *Machine Interpretation of Line Drawings*, MIT Press, Cambridge, Mass., 1986.

14. N. White and W. Whiteley, 'The algebraic geometry of stresses in frameworks', *SIAM Journal of Algebraic and Discrete Methods* 4 (481-511), 1983.

15. N. White and W. Whiteley, 'The algebraic geometry of motions in bar and body frameworks' *SIAM Journal of Algebraic and Discrete Methods,* **to appear**.

16.   W.-T. Wu, 'On the decision problem and the mechanization of theorem-proving in elementary geometry', *Contemporary Mathematics* **29**, 1984.

Discussion

(This paper was presented by Crapo.)

Davenport: Were you doing complex projective geometry or real projective geometry?

Crapo: Real—just a personal preference.

Davenport: In that case it's not clear to me how ideal theory *per se* will solve all your problems.

Crapo: No, I don't expect it will.
  Do you know whether in Macaulay or any other existing program you can find all the minimal prime ideals over an ideal?

Davenport: Yes, there is Scratchpad; the best person to talk to on this is Patrizia Gianni of the University of Pisa.   She has an algorithm for the minimal prime ideals in radicals.

Fisher: About ten or fifteen years ago there was a lot of work in AI on constraints in interpreting line drawings of polyhedral-type worlds. That's not a constructive, but rather an analytical type of geometry.

Crapo: Where you try to work out which line is hiding which by using sign rules and that sort of thing?   There's a new book which deals with that and with projective conditions, by Sugihara [13], which updates that work.

3 Robot motion planning

J. H. DAVENPORT

3.1 Introduction

We consider a space (typically of two or three dimensions) which contains a number of **forbidden regions**, and a **robot** which has to move in this space, and in particular to get from one point to another. In fact, it is not sufficient for it to be able to get from one point to another—we require it to get from one **situation** to another, where a situation consists of the location of all parts of the robot e.g. 'Point A here, the feet level and together, this edge facing north, the arms unfolded and pointing east ...' In addition, there may be other constraints, such as having to remain supported by the floor.

Throughout this talk we are interested in the **strategy** of the move, rather than in the **tactics**. In other words, we are interested in the question 'What routes are possible?', rather than 'Which is the best route?' or 'How close to this wall should I steer?' These questions are interesting in themselves, but demand a rather different knowledge of the world (and of the robot) than the strategy question. It is clear that the list of possible routes must be compiled *before* we can hope to answer any questions about the best route within it, and so the strategy question is, in some sense, the most fundamental. Having said that, a possible railway route from Winchester to London is via Southampton-Portsmouth-Le Havre-Paris-Calais-Dover, but one would not like this to be suggested as the first possibility. More prosaically, one would not wish to consider the solutions which involved the robot completely withdrawing from the areas of interest, going round the outside, and entering from the other side, unless all direct solutions had failed.

These strategy problems are 'intuitively easy' for human beings, who tend to solve them by experience and intuition. Typical examples include:

- The piano movers (or sofa movers, according to Howden [13]) who try to manoeuvre a large object through narrow doors, up steps etc.

- Car parking, or more particularly the backing of articulated lorries, where the system of constraints on the wheels is particularly severe.

- Robot motion planning proper: e.g. welding inside a nuclear reactor.

It is normal to assume that the robot, obstacles and constraints are given by a polynomial equations; in practice this does not seem to restrict generality. Another assumption generally made is that the robot consists of a finite (generally small) number of rigid parts. While quite realistic for robots, this rules out the modelling of human beings. Circles are defined by equations such as $x^2 + y^2 = 1$, so that there is no need to introduce trigonometric functions. This sort of constraint is particularly common, since it states that two points are the same distance apart, and explains why non-linear equations occur even in apparently linear problems. The forbidden regions are assumed to be defined by polynomials which in the simple case are positive in 'acceptable' regions, zero on the boundary, and negative in the forbidden part. In more complex cases, we may require logical combinations of such inequalities; for example a square pillar could be defined by

$$x \geq 0 \cap x \leq 1 \cap y \geq 1 \cap y \leq 1.$$

Let us introduce various definitions that will be of use in examining the complexity of the problem being solved. The **space dimension** d_S is the dimension of the region of Euclidean space in which we are interested—normally two or three. The **algebraic dimension** d_A of a problem (more precisely, of the formulation of a problem, since we could have inefficient formulations) is the number of variables needed to define the position of the robot unambiguously: i.e. the number of variables needed to define a situation. For the (not very realistic) case of a point robot, this is equal to the space dimension, for a robot whose situation is defined unambiguously by k points, the algebraic dimensions is k times the space dimension. The **geometric dimension** d_G of a problem is the number of 'degrees of freedom' present in the problem—more formally it is the dimension of the algebraic variety in a space of d_A dimensions, engendered by the qualities defining the position of the robot. The next section will give some examples of these definitions.

3.2   The simplest case

The simplest case is where the interesting region of space is two-dimensional (i.e. the space dimension is 2), the robot is a one-dimensional straight line and all boundaries of forbidden regions are defined by linear polynomials. This is certainly a simple case, but sufficient to illustrate the complexity. In this case, the position of the robot is determined by any two points on it; for simplicity we will take the two ends and call them (x, y) and (x', y'). Hence the algebraic dimension is 4. However if l is the length of the robot, then there is an equation connecting the two points, viz:

$$(x - x')^2 + (y - y')^2 = l^2.$$

Hence, the geometric dimension is 3, since we can choose x, y, and y' at will and then have at most two choices for y'. Of course, we may have zero, but that will only occur if x and x' are incompatible.

In this case, there is an algorithm due to Schwartz and Sharir for solving the robot motion planning problem. We give now a sketch of their algorithm; the details and the rather intricate arguments that justify it should be investigated in their paper [16]. We define a **wall** to be a part of the boundary of a forbidden region which is defined by a single linear equation. Hence the square pillar mentioned earlier is defined by four walls.

Essentially, the aim is to divide d_S-dimensional space into 'regions', such that the robot's situation is equivalent throughout the same 'region'. The word 'region' has been placed in inverted commas, since the situation is actually not so simple. In the special case we are considering, an orientation of the robot consists of a point (x, y) *and* an orientation of the other end (x', y') of the robot. Hence a 'region' must be more detailed than just a region of two-dimensional space—consider the example of a very narrow vertical corridor $0 \le x \le 1$ with a very long robot in it, when the position $x = 0.5, y = 0$ with $x' = 0.5, y' = 0.5$ is very different from the position when $x' = 0.5, y' = -2.0$.

Hence a region needs to have orientation information about the robot, as well as positional information. From any given point, number the walls clockwise as one encounters them, counting only those walls that are within 1 of the origin—we assume that the numbering is circular, so that wall 1 occurs immediately after wall n (the last wall). Each such wall will define (at most) two 'stops' on the position of the robot - one which prevents the robots from moving more clockwise, denoted k_c for the stop against wall k and one which prevents it from moving more anticlockwise, denoted k_a. Then a situation consists of a position, and an orientation which lies between two stops k_a and $(k + 1)_c$.

We therefore define a 'region' to be an area of space together with a range of permissible orientations, i.e. a number k such that all permissible orientations are between k_a and $(k + 1)_c$. Of course, we will need to know that the same walls are numbered throughout the area of space and that it is in fact this requirement that determines what a region looks like. Schwartz and Sharir [16] prove that there exists a finite set of curves (the **critical curves**) which divide the whole of two-dimensional space into areas such that the stops have a consistent numbering in each area. These curves can be thought of as resulting from the interaction of the robot and the walls, and fall into the following categories (we ignore various degenerate cases, and refer the reader to Schwartz and Sharir [16] for the detailed treatment):

- Straight lines resulting from the interaction of the robot with walls—these lines are parallel with the walls and a distance 1 from them.

- Arcs of circles resulting from the interaction of the robot with corners (where two walls meet)—these circles have radius 1 and are centred at the corner.

- More complex curves (**conchoids of Nicomedes**) which are the curves of degree 4 described by one end of the robot, while the other end is sliding along a wall and the middle of the robot is pivoting about a corner.

If there are n walls, then there are $O(n)$ curves of the first and second kinds, and $O(n^2)$ curves of type 3. These curves therefore divide space into at most $O(n^4)$ areas. This might lead one to think that there would be an immense complexity of the crossing rules from one region to another, but, since the degrees of all the curves involved are bounded by 4, there can be at most $O(n^4)$ crossings from one area of two-dimensional space to another. We still have to deal with the problem or orientation, but this only adds at most another factor of n, so that we have at most $O(n^5)$ edges in our graph.

The net result of the analysis is that there are a large number of regions (i.e. areas with orientations), which can be thought of as forming the vertices of a graph, with two regions being connected if they are neighbours. Some regions are 'allowed', others are 'forbidden' (since the robot would cross some wall if it were in them). The set of allowed regions forms a sub-graph of the set of all regions. The motion planning problem can now be stated simply: it is possible to pass from one situation to another if, and only if, the corresponding regions are connected in the graph of allowable regions.

Give the graph, determining this information is straightforward. Indeed, the usual algorithm [7] will produce the paths in 'shortest-first' order, which corresponds to the requirement stated in the introduction for a lazy evaluation of the list of possible strategies.

Having said that the algorithm is comparatively simple and requires a polynomial number of operations, it is necessary to point out that all is not rosy in the garden. Firstly, there is no guarantee that the operations involved are easy to perform. Indeed, the mere determination of the equations for the curves may be quite difficult, and determining the intersections may have a high computational complexity, since the intersection of two conchoids of Nicomedes requires one to isolate the real roots of a polynomial of degree 16 [4]. These root determinations need to be carried out with substantial accuracy, since the difference between deciding that there is an intersection and that there is not affects the structure of the graph, and hence all deductions about reachability.

As far as the present author knows, this algorithm has never actually been implemented; it would be interesting to do so!

3.3   Simple generalizations

The above algorithm can be generalized to the case of two-dimensional robots (still with straight-line sides) [16]. The argument is not very different and still proceeds by dividing the space up into a number of areas and defining a region as an area coupled with a range of orientations between two stops. The various curves required to divide up space are now somewhat more complex (and more numerous) but the algorithm is still polynomial in the number of walls (assuming that the complexity of the robot is fixed).

In a different direction, the algorithm has been generalized [20] to the case of a **spider**, a robot consisting of one point to which are attached the ends of k different one-dimensional arms. Hence the case considered in the previous section can be regarded as $k = 1$. The argument is similar to that of the previous case, except that we now need to consider k orientations, instead of one. The complexity is then $O(n^{k+4})$. Again, we have polynomial complexity in the number of walls, assuming that the complexity of the robot is fixed.

These methods have been generalized to a very special case of multi-body motion, that of circular objects moving amidst polygonal barriers in two dimensions [18]. There are related generalizations to the problems of two-dimensional linkages [10,12]. In these cases, though the object may have a large number of components, the interconnections are fairly simple—each item to the next. The problems are similar, because the two-dimensional linkages are not allowed to cross themselves. The only three-dimensional generalization known to the author is that of Schwartz and Sharir [19].

3.4 The polynomial case

Let us now consider a far more general case, that of an arbitrary, but polynomially-defined, robot moving in three dimensions among polynomially-defined obstacles. In this case, there is, in principle, an algorithm which will resolve the problem, due to Schwartz and Sharir [17]. This algorithm relies on the very general technique of **cylindrical algebraic decomposition** [1,2,3,4].

Space does not permit a complete description of this complex algorithm, so we will content ourselves with a description of the behaviour of the algorithm. It allows one to take an arbitrary part of n-dimensional space defined by a set of polynomial equations and inequalities (formally, a **semi-algebraic variety**) and decompose it into connected regions (**cells**), and to investigate the adjacency relations between the cells.   Once one had determined all the adjacency relationships, one has a graph just as in the very simple case studies, and the paths in the graph from the starting situation to the ending situation correspond to the possible strategies.

The caveat in principle mentioned at the start of this section is most certainly required. Let us consider a very simple example, that would be solvable by the basic two-dimensional algorithm. Take a corridor with a right-angled bend, formed by four lines

$$x = 0 \cap y \leq 0$$
$$x = 1 \cap y \leq -1$$
$$y = 0 \cap x \geq 0$$
$$y = -1 \cap x \geq 1,$$

and attempt to pass a line of length 3 round the right-angled bend. This attempt will fail, since $3 > 2\sqrt{2}$, but it is instructive to observe how expensive the failure is.

The space of feasible configuration of the ladder alone is defined by $(x - x')^2 + (y - y')^2 = 9$. There are then four constraints which says that the ladder must not cross any of the four walls bounding the corridors.   Consider the wall $y = 0 \cap x \leq 0$. Then there are two inequalities which must both be satisfied for there to be an intersection between the ladder and the wall, which can be written as $yy' \leq 0$ (the two ends of the ladder must be on opposite sides of the infinite line to which the wall belongs) and $x + y(x' - x)/(y - y') < 0$. The latter is not a polynomial, and must be recast as $x(y - y')^2 + y(x' - x)(y - y') < 0$. It is worth noting that this is already an equation of total degree 3, despite the fact that it deals with the intersection of line segments.

There are a total of four such pairs of equations, so the total system of polynomials looks like:

$$(x - x') + (y - y') - 9$$
$$yy'$$
$$x(y - y')^2 + y(x' - x)(y - y')$$
$$(y - 1)(y' - 1)$$
$$(x + 1)(y - y')^2 + (y - 1)(y - y')$$
$$xx'$$
$$y(x - x')^2 + x(y' - y)(x - x')$$
$$(x + 1)(x' + 1)$$
$$(y - 1)(x - x')^2 + (x + 1)(y' - y)(x - x').$$

Davenport [5] describes the reduction of these equations by means of the cylindrical algebraic decomposition algorithm. The projection into three variables (eliminating x) gives 32 polynomials: six contents of the original non-primitive polynomials, six partial derivatives, four non-constant coefficients, one discriminant and 15 resultants. There are actually only 10 distinct polynomials involving y (the next variable to be eliminated), and these are:

$$yx' + y - x' - x'$$
$$y - y'$$
$$y$$
$$yx' + y'$$
$$y - 1$$
$$y^2 - 2yy' + x'^2 + y'^2 - 9$$
$$y^2 - 2yy' + x'^2 + 2x' + y'^2 - 8$$
$$y^2 x'^2 + y^2 y'^2 - 2yx'^2 y' - 2yy'^3 + x'^2 y'^2 + y'^4 - 9y'^2$$
$$y^2 - 2yy' + y'^2 - 9$$
$$y^2(x'^2 + 2x' + y'^2 - 2y' + 2) + y(- 2x'^2 y' - 4x'y' - 2y'^3 + 4y'^2 - 4y')$$
$$\quad + x'^2 y'^2 + 2x'y'^2 + y'^4 - 2y'^3 - 7y'^2 + 18y' - 9.$$

Projecting these polynomials into two dimensions gives us 79 polynomials in the two variables. They can be reduced to 24 non-trivial ones, the largest being

$$x'^4 y'^4 + 4x'^3 y'^2 + x'^2 y'^4 - 2x'^2 y'^3 - 2x'^2 + 18x'^2 y' - 9x'^2$$
$$\quad + 2x'y' - 4x'y'^3 + 6x'y'^2 + y'^4 - 2y'^3 + 2y'^2.$$

When we project into one variable y', we have to compute 19 discriminants and 276 resultants which, using REDUCE 3.0 on a DEC 2060, takes 191.4 seconds. These can be simplified to a total of 184 polynomials of total degree 801. Of these, 27 are linear and the rest have even degrees ranging from 2 to 18. The polynomial of degree 18 is

$$8y'^{18} - 72y'^{17} + 1632y'^{15} + 6400y'^{14} + 10528y'^{13}$$
$$\quad + 55332y'^{12} - 373384y'^{11} + 789824y'^{10} - 528340y'^9 - 359993y'^8$$
$$\quad + 13488560y'^7 - 15989896y'^6 + 2615524y'^5 + 16276624y'^4 - 22155768y'^3$$
$$\quad + 13667292y'^2 - 4251528y' + 531441.$$

This polynomial is, in fact, irreducible, which means that it is almost certainly spurious, since we remarked above that the most complex equation required by the alternative method is the intersection of conchoids of Nicomedes, which has degree 16, not 18.

These polynomials have a total of 375 real roots, and a continuation of the computation would have required one to arrange all these roots in ascending order along the real line and then analyse the behaviour of the bivariate polynomials above each of them, and so on. Unfortunately, all the author's attempts to continue with this example have failed [5], though recent research [8] gives some slight hope.

Cylindrical algebraic decomposition is inherently a very expensive process. If we have n polynomials of degree d in r variables, whose numeric coefficients are bounded by 2^h, then the total computation time for the algorithm is bounded by

$$n^{2^r}(2d)^{r^{2^{r+s}}} h^3,$$

which is doubly exponential in r [4]. This doubly exponential nature is inherent, since there are examples on which the output of a cylindrical algebraic decomposition in r variables requires at least $2^{2^{r/5}}$ different regions [6]. Further results in complexity have been proved by Hopcroft et al. [11] and Reif [15].

3.5 Conclusions

Robot motion planning is a subject in which the capabilities of the extant algorithms are clearly far removed from the practical problems that we would wish to address. There is much more scope for algorithmic research, both directly in robot motion planning algorithms, and the the sub-algorithms such as real root isolation and the description of cuirves and surfaces, and their intersections. This point has been made by Hopcroft [9]. Equally, more practical experience with these algorithms needs to gained, since it may well be that there are heuristics which would speed up 'common' cases. Davenport [5] makes about the only report of a practical investigation.

Recently an alternative approach has been described by O'Dunlaing et al. [14], though it is too early to assess its computational significance. It still seems to use the same basic algebraic machinery as the algorithms we have mentioned and therefore to benefit from the sort of algorithmic improvement we have mentioned.

References

1. D.S. Arnon, G.E. Collins and S. McCallum, 'Cylindrical algebraic decomposition I: the basic algorithm', *SIAM Journal of Computation* **13** (865-877), 1984.

2. D.S. Arnon, G.E. Collins and S. McCallum, 'Cylindrical algebraic decomposition II: an adjacent algorithm for the plane', *SIAM Journal of Computation* **13** (878-889), 1984.

3. G.E. Collins, 'Quantifier elimination for real closed fields by cylindrical algebraic decomposition', Proceedings of the 2nd GI Conference on Automata Theory and Formal Languages (Springer Lecture Notes in Computer Science 33, Berlin) MR55--#771 (134-183), 1975.

4. J.H. Davenport, 'Computer algebra for cylindrical algebraic decomposition', KTH Stockholm NADA Report TRITA-NA-8511, September 1985.

5. J.H. Davenport, 'On a "piano movers" problem', *ACM SIGSAM Bulletin* **20**,1&2 (15-17), 1986.

6. J.H. Davenport and J. Heintz, 'Real quantifier elimination is doubly exponential', **to appear in** *Journal of Symbolic Computation* **3**, 1987.

7. E.W. Dijkstra, 'A note on two problems in connexion with graphics', *Numerical Mathematics* **1** (269-271), 1959.

8. J. Hollman and J. Schubert, 'Computer algebra for root isolation', KTH Stockholm NADA Report TRITA-NA-8663, 1986.

9. J.E. Hopcroft, 'The impact of robotics on computer science', *Communications of the ACM* **29** (486-498), 1986.

10. J. Hopcroft, D. Joseph and S. Whitesides, 'Movement problems for 2-dimensional linkages', *SIAM Journal of Computation* **13** (610-629), 1984.

11. J. Hopcroft, D. Joseph and M. Sharir, 'On the complexity of motion planning for multiple independent objects; PSPACE-harness on the "warehouseman" problem', *International Journal of Robotics Research* **3**,4 (76-88), 1984.

12. J. Hopcroft, D. Joseph and S. Whitesides, 'On the movement of robot arms in 2-dimensional bounded regions', *SIAM Journal of Computation* **14** (315-333), 1985.

13. W.E. Howden, 'The sofa problem', *Computer Journal* **11** (299-301), 1968.

14. C. O'Dunlaing, M. Sharir and C.K. Yap, 'Generalized Voronoi diagrams for moving a ladder I: topological analysis', *Communications in Pure and Applied Mathematics* **39** (423-483), 1986.

15. J. Rief, 'Complexity of the mover's problem and generalizations', Proceedings of the 20th Symposium on the Foundations of Computer Science (421-427), 1979.

16. J.T. Schwartz and M. Sharir, 'On a "piano-movers" problem I. The case of a two-dimensional rigid polygonal body moving amidst polygonal barriers', *Communications in Pure and Applied Mathematics* **36** (345-398), 1983.

17. J.T. Schwartz and M. Sharir, 'On the "piano-movers" problem II. General techniques for computing topological properties of real algebraic manifolds', *Advances in Applied Mathematics* 4 (298-351), 1983.
18. J.T. Schwartz and M. Sharir, 'On the "piano-movers" problem III. Coordinating the motion of several independent bodies: the special case of circular bodies moving amidst polygonal barriers', *International Journal of Robot Research* 2 (46-75), 1983.
19. J.T. Schwartz and M. Sharir, 'On the "piano-movers" problem IV. The case of a rod moving in three-dimensional space amidst polyhedral obstacles', *Communications in Pure and Applied Mathematics* 37 (815-848), 1984.
20. M. Sharir and E. Ariel-Shaffi, 'On the "piano-movers" problem V. Various decomposable two-dimensional motion-planning problems', *Communications in Pure and Applied Mathematics* 37 (479-493), 1984.

Discussion

Prior: Is the curve of Nichomedes a polynomial, or a more complex form?

Davenport: It's a polynomial of degree 4: it's not particularly difficult to derive the equation.

Sabin: You were talking about connectivity of regions. Could you achieve the same result by arguing about the connectivity of region edges?

Davenport: That is how you actually work out which regions are connected, but you must also consider changes of allowable orientation between regions. As you cross an edge, the set of available orientations changes.

Sabin: Yes, but you could also consider yourself sliding along the region edges. Where they meet the boundaries will then constrain the search for a solution.

Davenport: How do you know which edge to slide along?

Sabin: It's a graph-theoretic connectivity problem, isn't it?

Davenport: The edges are the dual structure to the graph of spaces, but we need the orientations attached to the regions. I don't see how to map that orientation information from the space graph onto the edge graph. We proceed by constructing the edge graph, derive its dual, and then label that with orientations. You could perhaps pull the orientations back onto the edge graph, but that would probably make life harder.

Sabin: But surely one can construct certain arguments about the connectivity of a manifold, in terms of the connectivity of its boundary?

Davenport: That would be fine if the space were totally partitioned, so that the problem could be solved irrespective of orientation.

Martin: There is one thing you could do; in your example, instead of taking what is essentially a vector sum of the wall with the object, use one end of the line and its angle. If you did a vector sum with that, then you'd end up with a three-dimensional shape, and the problem would boil down to deciding whether the vector sum consisted of one connected region or disconnected pieces. That would seem to avoid the necessity for solving such high orders of polynomial. You are only adding circles to straight lines, so you can't end up with very high-order polynomials, can you?

Davenport: In the three-dimensional manifold, no, but how are you going to decide whether the manifold is connected or not?

Sabin: Find all the pieces connected to one end, and then determine whether the transitive closure of them includes the other end.

Davenport: How do you do the transitive closure of components? You break up your three-dimensional shape into simple cells.

Sabin: I think I can conduct the argument by crawling over the surface of this three-dimensional manifold, which brings it down to two dimensions.

Davenport: We must discuss this further. The general Schwartz and Sharir method looks at the entire manifold, and this resulted in the equation of degree 810 in the example. So looking at the manifold itself is not attractive; that's what I was doing with the general method.

Woodwark: If someone asked me to take one of these robots with lots of arms down to the foyer of this building, then I might be inclined to program a computer and wait a week for it to tell me how to do it. But if someone asked me to take a small suitcase down to the foyer, then I'd set off without further ado. I've seen something of this path-planning work before, and it seems to me that what is lacking from a practical point of view is ways of making the safe approximations that a human makes: knowing that it is possible to take a suitcase to the foyer. In planning for real robots moving around factories, that is the sort of computation that will be required 99% of the time. I'm not saying that doesn't involve any algebra; I think you're probably doing certain processes mentally that do correspond to simple algebraic manipulations. Have you any insight into how we can cut down the size of problems that we have reason to expect to be trivial? Creating all these regions between here and the foyer and so on for my suitcase problem would produce a solution, but it would be a sledgehammer to crack a nut.

Davenport: I've had a couple of ideas about this. Looking at the complexity theory, I said that anything guaranteed to produce all strategies is bound to have this problem. You don't want all strategies. If I 'phone up the railway station and find out how to get from Winchester to Waterloo, they're more likely to think of a direct train, rather than a route via Southampton, Portsmouth, Le Havre, Calais, and Dover, although that's equally feasible! To avoid doing all the work, start from the initial position and set out to explore the graph between that and the goal using what is essentially

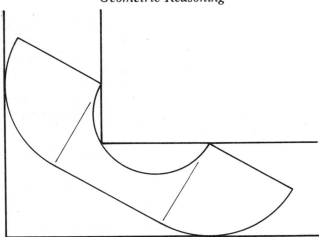

Fig. 3.1   A conjectured maximal area shape that will go round a corner in a
corridor. Its overall length is $(2 + 4/\pi)$ times its width, which is the
same as that of the corridor.

Dijkstra'a inkblot algorithm; then you only actually build the parts you need
as you go along. I think that approach will remove most of the complexity
for simple problems, because you won't look at parts of the problem you don't
need to solve: technically, it's called lazy evaluation.

Fisher: Would you care to comment on articulated robots—the non-rigid case?
Can the algorithm be generalized?

Davenport: Articulated robots are generally still piecewise rigid.

Fisher: Yes, jointed.

Davenport: It'll solve those problems just as well.   The whole thing
generalizes: you just end up with a set of critical curves for each component.

Fisher: So you intersect the paths, in the sense that there's a constraint linking
the two rigid sub-components together.

Davenport: Each component has a reduced number of degrees of freedom;
what I call the feasible algebraic dimension is lower, but the methodology still
deals with it. (Except to the extent that, as I've already said, the methodology,
while valid, is infeasible in practice!)

Crapo: Is there any description of a family of largest possible objects that can
go round a corner?

Davenport: That's a much harder one. In the example I gave I know how
to find out what the largest one-dimensional polygon is. A still unsolved
problem is to determine the largest object in area to go round the corner in
my example. The best solution to date (see Fig 3.1) is a somewhat eccentric
one, which has the largest known area $\frac{\pi}{2} + \frac{2}{\pi}$, but that hasn't been proved
to be maximal.

Bowyer: From considerations of elegance, I expect that *is* the maximal solution!

Davenport: It's a strong conjecture! Note that the semi-circular cut-out in the shape *increases* the total area that can go round. Technically speaking, all the problems I mentioned so far are first order problems in the theory of real closed fields, but this is a second-order one.

Crapo: In the two-dimensional case, is there no short cut to the result $2\sqrt{2}$? That length could then be compared with a candidate line to determine whether *that* could get around the corner.

Davenport: I don't know of a good way to do that. As soon as I try to find out that the answer is $2\sqrt{2}$, I'm working in a higher-dimensional space. Essentially, the length of the line must be added as an extra parameter, and that makes everything more complicated.

4 Programming interactions by constraints

S. J. P. TODD

4.1 Introduction

A typical scenario in interactive graphics is that a model is defined by the application and sent as a picture to the graphics system. The picture is sent with certain parameters still not defined: typically an overall viewing transform, or a transform that describes how one part of the model may be moved relative to the rest. The graphics system then monitors interactive devices such as dials and mice, and modifies the picture parameters appropriately.

A major human factors concern is what we mean by 'appropriately'. Certain principles are known. For example, 'kinesthetic correspondence' [5] states that when a device is moved in a particular direction, the object it is controlling should move in the same direction. Generally, we hope to connect the device to the object in a 'natural' manner.

Unfortunately, a mechanism that appears natural to someone sitting in front of a display often requires a complex program to read the input devices and change the picture parameters. We call this the 'interaction definition' program. The complexity makes experimentation with different interaction techniques very expensive in programming time and limits the amount of experimentation that can be performed.

This paper proposes a program generator mechanism for defining interactions in terms of constraints. The interaction definition is made as a set of geometric statements. These statements are converted into an algebraic form, and compiled (solved) by the computer algebra into a conventional program. This program is then inserted into the graphics system between the devices and the picture parameters.

Constraints have been used before in graphics systems: in font definition [4], typesetting and picture layout [7], and robotics and solid modelling [6]. Our proposal is to extend their work into the area of defining interactions.

Geometric Reasoning

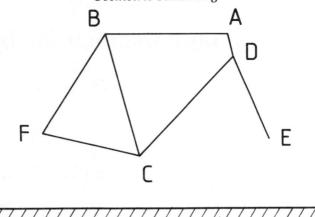

Fig. 4.1 A diagrammatic two-dimensional bicycle. The quadrilateral **ABCD** is the main frame. The line **DE** represents the front forks. The triangle **BCF** represents the rear forks. The wheels are centred on **E** and **F**, the pedals on **C**. The ground is at $y = 0$.

This paper is in two major parts; the first gives an overview of the system; the second describes the computer algebra problems encountered, and how these constrain the use of such a system in practice. We have built a simple prototype and we use this in examples.

4.2 Overview

The system is in three parts.

- The geometry converter: this takes the interaction definition in geometry form and converts it to algebraic form.

- The equation solver: this takes the algebraic form, and converts it to a conventional program. The inputs to the program are the interaction device values. The outputs are the picture parameters.

- The run-time system: this monitors the interaction devices, runs the program generated by the equations solver, and feeds the picture parameters to the graphics system.

We give an example of interaction with a stylized bicycle, as shown in Fig. 4.1. Specifying the following variables permits a picture of the bicycle to be realized:

Crossbar horizontal at height $h = y_A = y_B$.
Wheel diameter $w = y_F = y_E$.

Clearance for pedals: $p = y_C$.
Wheelbase: $l = x_E - x_F$.

This example uses geometric concepts of *LENGTH* of a line, and *DIRECTION* of a line from horizontal. We define some values as constants:

$$DIRECTION\ (\mathbf{B}, \mathbf{C}) = -72°$$
$$DIRECTION\ (\mathbf{A}, \mathbf{D}) = -72°$$
$$LENGTH\ (\mathbf{B}, \mathbf{C}) = 23''$$
$$LENGTH\ (\mathbf{D}, \mathbf{A}) = 4''.$$

We connect other values to system interaction devices, using the functions *VALUATOR* and *CURSOR*:

$$LENGTH\ (\mathbf{F}, \mathbf{C}) = VALUATOR\ (1, 16)$$
$$\mathbf{A} = CURSOR\ (1,\ (24, 1))$$
$$\mathbf{B} = CURSOR\ (2,\ (1, 1)).$$

The second parameter of these functions is an initial value before interaction. For our simple system we use all devices as relative, though absolute devices fit easily into the design.

The geometry converter passes simple statements such as $w = y_E = y_F$ directly to the algebra system. Point declarations are used to define a pair of variables; *POINT* (**A**) generates variables x_A and y_A. Geometric concepts are expanded, so *LENGTH* (**B**, **C**) $= 23''$ is sent to the algebra system as $(x_B - x_C)^2 + (y_B - y_C)^2 = 23^2$. *VALUATOR* and *CURSOR* are replaced by variables *VALUATOR1* and so on.

The algebra system solves the equations it has been presented to generate a program. *VALUATOR1* etc. are the inputs to the program; x_A etc. are the outputs. The algebra system may encounter several cases.

- The user has not specified enough constraints for the system to be solved. This is reported to the user.

- There are enough constraints, but the equation solver is not powerful enough to cope. This all too common case is discussed more fully below.

- The user has specified too many constraints. Some are used to generate a solution. The remainder are added to the program as constraints to be checked as the program is run. Inequality constraints, such as the wheel must not pass through the frame, can only be used as run-time checks.

In our experimental system, the algebra system usually outputs a program in a p-code, which is interpreted at run time. It can also create a BASIC program.

There are two special features of the run-time program, used to disambiguate square roots, or the solutions of higher-order polynomials.

- The previous two values of each variable are stored. These are used to provide an 'inertia' effect in the system. For example, an interaction equivalent to the user pushing a wheel round by pumping a piston does make the wheel go round rather than oscillate.

- A re-try mechanism is provided at each point of ambiguity. If a constraint fails, the re-try is invoked. This permits a constraint such as the non-interaction of wheel and frame to force a correct configuration.

The bicycle example is used as it is easy to explain. In a typical interaction set-up, the picture is passed directly to the graphics system parameterized by transforms. The geometry system is only told about those parts of the picture needed to define the interaction, and how they relate to the transforms. The processing is similar, and the final product is a program that takes the devices as inputs, and produces the transform values as output.

4.3 The algebra system

The conversion of geometry to algebra is straightforward. Solving the algebra generated is much more difficult. We were unable to find a package that could do the required work, and were forced to build our own equation solver. We do not go into detail here, but list a few reasons why the current computer algebra systems are not suitable for our needs.

- Most systems are designed to be led interactively by a user. They have low-level processing power for differentiation and so on, but do not include higher-level strategies for solving a complete set of equations.

- Most systems are designed as stand-alone systems, and are not suitable for interaction with other applications.

- Computer geometry generates simultaneous non-linear equations, which are outside the domain of most systems. Computer geometry is now a subject in its own right, used especially in robotics [e.g. 1,2]. and a technique called the **Gröbner-basis** algorithm exists for simultaneous non-linear equations.

All of the above points are being addressed by the new generation of computer algebra systems, such as SCRATCHPAD II [3].

When the algebra system fails we have a problem reporting the error back through the geometry system in a way that is meaningful to the user. This is similar to the problem often encountered with macro front ends. Also, two interaction definitions that appear similar to the user may look very different algebraically. This makes it difficult for the user to predict exactly when the system is likely to fail.

These last problems mean that interaction definition programming remains a task for the specialist.

4.4 Conclusions

We have outlined a system for interaction definition using constraints. It is intended as a tool for human factors experimentation. The user specifies an interaction as a set of geometric constraints. These are converted by the system to algebraic equations, which are solved to provide an executable program taking interaction device values as input and producing picture parameters as output.

This method speeds up experimentation with different interaction techniques. We have built a prototype system that demonstrates the idea to be feasible and valuable. The idea is limited by the state of the art of computer geometry systems; but these systems are improving rapidly. The method is good at simplifying the task of the specialist programmer, but the complexities of the problem still preclude the generation of interaction programs by end users.

References

1. J.H. Davenport, 'Robot motion planning', in *Geometric Reasoning*, (Proceedings of a conference held at the IBM UK Scientific Centre, December 1986), OUP, 1989.

2. A. Bowyer, J. Davenport, P. Milne, Julian Padget and A.F. Wallis, 'A geometric algebra system', in *Geometric Reasoning*, (Proceedings of a conference held at the IBM UK Scientific Centre, December 1986), OUP, 1989.

3. R.D. Jenks, R.S. Sutor, and S.M. Watt, 'Scratchpad II: an abstract datatype system for mathematical computation', Lecture Notes in Computer Science 296, (Proceedings of an International Symposium on Trends in Computer Algebra, Bad Neuenahr, May 1987, R. Janssen, ed.) Springer Verlag (12-37), 1987.

4.   D.E. Knuth, 'TEX and Metafont, new directions in typesetting', Digital Press, Educational Services, Digital Equipment Corporation, 1979.

5.   J.S. Lipscom, 'Three-dimensional cues for a molecular computer graphics system', PhD Dissertation, University of North Carolina, Chapel Hill, North Carolina, 1979.

6.   R.K. Popplestone, A.P. Ambler and I.M. Bellos, 'An interpreter for a language for describing assemblies', *Artificial Intelligence* 14 (79-107), 1980.

7.   C.J. VanWyck, 'A graphics typesetting language', *SIGPLAN Notices,* 16,6 (Proceedings of an ACM SIGPLAN SIGOA Symposium on Text Manipulation, Portland, Oregon) (99-107), June 1981.

Discussion

Crapo: How do you decide in advance on a what is likely to be a good set of free variables?

Todd: Using the intuition of the person doing the human factors experiment or whatever: when one starts to use a system one often gets the feeling that the knobs available aren't doing what you want; sometimes things go haywire. If you speak to somebody who has used a system for half a day or so they say 'It would be so easy if I could just pick up a point and move it in such-and-such a way'. You could conduct complicated experiments, but I think it's better to rely on the intuition of the person using the system. The trouble is, when someone's used a system for too long, they say 'What I want is these knobs', because they've got used to them.

Crapo: The reason I asked the question is that you're in one of the areas that interest me the most, which is the special position problem. That will show practically by locking, you'll be in a position and you'll change something else a little too far, and you'll get locked in position because you've passed through a singularity with a branching.

Todd: Essentially, I ignore all the problems about branching, except for square roots, but that's not really a branching problem; it's really a multi-algebraic problem. I felt that if I ever get far enough for singularities to be a problem I'll be jolly pleased. The choosing of fixed and free variables is back to human factors—to the person using the system, so it's different from the singularity problem: initially, anyway.

Crapo: But even knowing how many free variables there are is a question of special positions.

Todd: That's true. I assumed the whole time that I was in general positions, although it wasn't really true.

Geisow: How does you work compare with some of the other constraint-based systems, especially Thinglab, Borning's work at Xerox [8]?

Todd: I think the main difference is in this two-phase approach.

Geisow: I understand that Borning compiles Smalltalk methods from a user description, but I'm not sure about this.

Todd: I'm not sufficiently familiar with that work.

Stobart: In RAPT, the University of Edinburgh assembly system, their first approach was to use symbolic algebra to manipulate spatial relations, but they found that a better and faster approach was to use what they call the cycle finder, which took standard solutions through the graph that they were able to form. It solved the majority of cases very much faster than the original equation solver; the improvement was many orders of magnitude. Would you consider that technique appropriate to your sorts of problem?

Todd: That was what I was saying about 'ignoring' the algebra. It didn't seem to me that their approach was very general; but possibly, while it lacked generality they were able to solve interesting problems. I did look at their approach, but only after I'd more or less finished this work anyway—perhaps a little late! My feeling is that it would not quite be general enough, but perhaps I don't appreciate how much work they've put into it.

Sabin: As I recollect, the Ivan and Sutherland Sketchpad system used a comparable strategy; it separated and then lumped together linear constraints, and then put further effort into the quadratic and higher order constraints.

Todd: But they did not use that strategy within the context of a general computer algebra system.

Sabin: No, of course that was one of the first constraints systems.

Martin: One thing that it seems that you could do here, considering that you're updating your picture every thirtieth of a second, is to use a differential approach. Replace your constraints by a description in differential terms: your quadratics would become linear, and the ambiguity would also disappear, as you would know the previous position you were at.

Todd: I didn't think of that, but I did consider multi-dimensional Newton-Raphson for solving the quadratics I couldn't deal with algebraically. You are really saying the same thing in a more elegant way, I think.

Woodwark: But you'd get drifting, presumably, after a bit.

Martin: Perhaps, yes.

Fisher: Did you consider constraints including some sort of imprecision, in the sense that you may not know the exact distance between two points, but only that it lies within a particular range?

Todd: That's a horrible problem by itself; it's probably impossible to deal with it at the same time as interactions. But perhaps not: I know that other people dealing with solving *static* constraints have been trying to deal with that type of thing as well.   It's probably not appropriate with these real-time constraints.

Davenport: Just a comment on what you said about strategy. A degree-sorted Gröbner-basis approach will more or less do that strategy as it will peel off all the linears first. Modern Gröbner algorithms will also let you classify variables into different categories, so you can arrange to eliminate all the known ones first, but that's a very recent development in the theory.

Todd: I still haven't got a copy of Scratchpad, which makes implementing strategies like that rather difficult.

Additional reference

8.   A. Borning, 'The programming aspects of Thinglab, a constraint-oriented simulation laboratory', *ACM Transactions on Programming Languages and Systems* 3,4 (353-387) October 1981.

5 Connectionist models and geometric reasoning

F. FALLSIDE and L.-W. CHAN

5.1 Introduction

There is much interest in processing information for machine understanding: for example speech and language in order to improve human-computer interfaces and vision for robotics applications and medical diagnosis. Each of these fields is highly specialized but share a broad division into low and high levels of processing. Low levels tend to be numerical: for example acoustic processing in speech and image processing in vision. Higher levels tend to be symbolic: for example the syntactic constraints on phonemic units in speech or on word units in language or primitives in vision [3]. Progress in machine understanding is relatively uneven and the results are far below human performance.

Most progress has been made in the two levels separately, particularly in symbolic processing based on languages such as LISP and Prolog and relatively little success has been achieved so far in bridging the gap and establishing high-level knowledge from real, physical data.

This is one of the reasons for the growth of interest in connectionist models. Other reasons are:

- To an extent they appear to be capable of both low-level and high-level processing; they can both abstract primitives and manipulate them.

- They have 'biological plausibility'. As massively-interconnected networks they may be analogous, at least in structure, to processing by the brain.

- They are inherently parallel in nature and we can thus exploit advances which are taking place in parallel-processing technology.

Connectionist models for pattern processing date back to early days in AI and the perceptron network [2,5]. Perceptrons failed to prosper

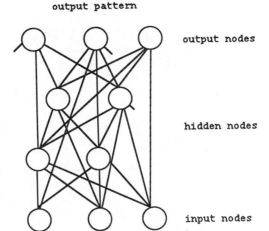

output pattern

output nodes

hidden nodes

input nodes

input pattern

Fig. 5.1 A connectionist network.

largely because there were no successful algorithms for training them. A resurgence of interest has taken place because successful training methods have recently been established. Research into them has blossomed as a result and an increasing number of types of network and training algorithms have been developed and are beginning to be applied to real, physical problems [7]. In this paper we briefly review a particular type of connectionist model, the back-propagation network, consider the use of such networks in geometric reasoning and then give a few early experimental results.

5.2 Connectionist models and the back-propagation network

Connectionist models

Connectionist models are a class of networks consisting of active non-linear nodes each of whose outputs is in general connected to the inputs of all the other nodes via weight coefficients, as shown in Fig. 5.1. There are three types of nodes: input nodes which are accessible for input patterns, output nodes which are accessible for output patterns, and 'hidden' nodes which are not externally accessible.

The network may be trained to recognize an input-output pattern relationship by adjusting all the weights to minimize the deviation of the actual output from the required output. When trained it will thus

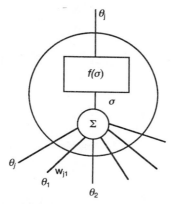

Fig. 5.2 The form of the nodes in a back-propagation network.

recognize test input patterns: for example, all instances of the same image or speech sound. Interestingly and importantly, if input data is incomplete it may 'generalize' by recognizing the complete data.

This type of training is termed **supervised training**, when the output pattern is specified by the user. Alternatively there is **unsupervised training**, when no output is specified and the input pattern simply defines a network weight set.

There are different types of network, for example stochastic networks, such as the Boltzmann machine which also has bi-directional links, and multi-layer perceptrons [7,8] which are deterministic. Networks are also characterized by their learning algorithms, for example **back-propagation** networks are multi-layer perceptrons which employ a particular least-squares energy algorithm.

Potentially such networks can be applied to a wide range of pattern processing, both numeric and symbolic. They have been applied to or proposed for a wide range of problems, for example:

- Simple logic functions such as XOR logic gates, binary encoders, binary addition and shift registers [7].

- Language processing: family trees and parsing [7].

- Speech processing: recognition [6] and synthesis [9].

- Vision processing: models of geometric reasoning and human vision processing [1].

We will now concentrate on the case of back-propagation networks and their application to geometric reasoning.

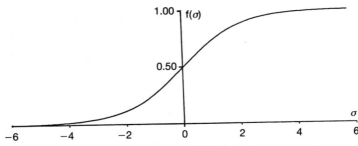

Fig. 5.3 The 'logistic' function.

Back-propagation networks

In this case the nodes are as shown in Fig. 5.2 with outputs as simple instantaneous non-linear functions of the total input. Thus θ_j the output of node j is given by:

$$\theta_j = f(\sigma_j) \tag{5.1}$$

$$\sigma_j = \Sigma_i w_{ji}\theta_i + \theta_j \tag{5.2}$$

where w_{ji} is the weight connecting the output of node i to an input of node j and θ_j is the bias.

The non-linear function commonly used is the 'logistic' function as shown in Fig. 5.3:

$$f(\sigma) = \frac{1}{1 + e^{-\sigma}}. \tag{5.3}$$

All the outputs are seen to lie in the range 0 to 1.

The network is trained for a particular input-output relationship by adjusting the weights w_{ji} and the bias terms θ_j of the whole network to minimize a network energy function. For this purpose, θ_j can be assumed to be connected to a constant value of unity, via an adjustable weight which is then treated like other weights.

The input nodes simply have an output consisting of the input data value.

The networks may be totally interconnected or divided into several layers: a layer of input nodes, one or more layers of hidden nodes and one layer of output nodes. Each layer is connected only to the one above it. An example is shown in Fig. 5.4.

Such networks are trained by a two-pass small-variation procedure. First, for a given input pattern (starting from random initial weight values) the output of each node is calculated from Equations 5.1 and 5.2 (which may be carried out in parallel), to the output nodes.

To establish the output values, say θ_{Oi} these are compared with the specified or target output values θ_{Ti} and errors for the output nodes:

$$e_i = \theta_{Oi} - \theta_{Ti} \tag{5.4}$$

are established and used in scaled form

$$\delta_i = f'(\sigma_i)(\theta_{Oi} - \theta_{Ti}). \tag{5.5}$$

Errors are now propagated back through the network using the expression

$$\nabla_j = f'(\sigma_j)\Sigma\delta_k w_{jk}, \tag{5.6}$$

where the scaled error for node j is connected to nodes k by weights w_{jk}. It can then be shown that for a network error function $E = \frac{1}{2}\Sigma_i(\theta_{Oi} - \theta_{Ti})^2$ summed over all patterns, a steepest descent solution $\Delta w_{ij}\alpha - \partial E/\partial w_{ij}$ provides

$$\Delta w_{ij} = \eta\delta_j\theta_i, \tag{5.7}$$

where η is known as the learning rate. (Again many of these calculations can be carried out in parallel.) The weights are now updated and the two passes are again repeated. These learning cycles are continued until the error function, E, for a given set of input patterns has decreased to a small value. For even quite small networks many learning cycles are required, often $\simeq 10^3$. This is one of the reasons why such networks are computationally very slow when carried out on conventional computers. Some examples of this are given later. This demonstrates the need to exploit parallel processing for this class of problem.

Often the learning equation is modified by a momentum term such that

$$\Delta w_{ij} = \eta\delta_j\theta(n + 1) + \varepsilon\Delta_{ij}(n). \tag{5.8}$$

5.3 Connectionist models and geometrical transformation

Over the last few years a number of studies have been made of a connectionist approach to object location, object recognition and navigation [1]. These started with two-dimensional objects and have now been extended to three-dimensional objects.

The basis for this is to solve a translational problem. For the two-dimensional case each edge section of a polygonal shape can be described by four parameters: horizontal position, vertical position,

output pattern

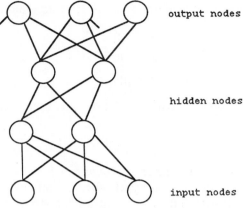

input pattern

Fig. 5.4 A layered network.

length and orientation, which for a polygon X can be described as $X = (x, y, l, \theta)$. Suppose we have a different view of the same polygon $X' = (x', y', l', \theta')$ then the two views will be related by the transformation $X' = TX$, where the transformation parameters can be described as the changes $T = (\Delta x, \Delta y, s, \Delta \theta)$. We see algebraically that:

$$\Delta\theta = \theta - \theta'$$

$$s = l/l'$$

$$\Delta x = x - x's \cos \theta - y's \sin \theta$$

$$\Delta y = y + x's \sin \theta - y's \cos \theta.$$

Suppose now we set up three networks: one for each X, X' and T as shown in Fig. 5.5.

Interconnecting them appropriately the networks will attempt to set up transformations between all the edges of X and those of X'. The 'answer' is given by the most polar single transformation in the scenes, which because of the rigidity is the correct one. It is shown, using a particular type of connectionist model (Hopfield network) that a cumulative process can be initiated to find the transformation at an energy minimum.

It has also been demonstrated that the method allows a particular object to be located in a cluttered scene because of rigidity. In our approach we are attempting to solve a similar problem using a CAD database of reference objects and back-propagation networks to establish the transformation from vision sensor images.

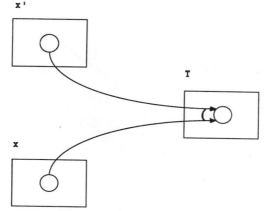

Fig. 5.5 Image transformation (after Ballard [1]).

5.4   Results for a particular pattern processing problem

The problem

The general problem we are approaching here is geometric reasoning about the recognition of solid objects and their orientation and location from two-dimensional images of physical objects or from solid models using a back-propagation network.   Our aim is to provide both an alternative to rule-based approaches which we are pursuing for a robotic assembly application [4,10] and as a parallel to human mental models of geometric reasoning.   We present some simple early results towards these ambitious aims.

The connectionist model

These results were obtained from variants of the particular form of back-propagation network shown in Fig. 5.6.   The input patterns are essentially two-dimensional binary wire-frame images.   As a result the input nodes are taken as lying in a grid or input plane.   These are connected to a single layer of hidden units, again taken as lying in a plane and finally a plane of output units.

The back-propagation algorithm already discussed was used with some variants.   The learning rate coefficients η and ε of Equation 5.8 are not constant but taken as functions of the previous learning path in weight-space.   This has been found to reduce the number of learning cycles required.

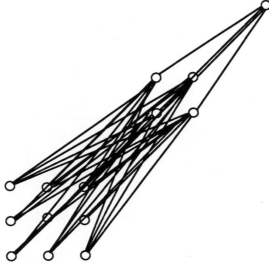

Fig. 5.6. A network with a plane of inputs.

Results for idealized data

The input patterns were obtained from views of a single solid object modelled by CATIA. For this 16×16 input nodes, 4×4 hidden nodes and 4×2 output nodes were used. The presence of an input point corresponds to an input node value of 1 and its absence to 0. Supervised training was used with particular value of θ_T for each view of the same object.

The aim of the experiment is to attempt to recognize different views of the same object, which give us some information about its orientation. The network is first trained with the eight training views to establish the network weights which minimize the error function. Objects are then recognized by applying test views to the network.

The results for eight equatorial views differing by about 45° appear in Fig. 5.7. In each case, the target output matrix θ_t, a training view, a test view and the resulting output matrix θ_O, are shown. The test views are slightly rotated versions of the training views in each case.

We see that network has been successfully trained to recognize a view and to discriminate between the eight different views.

The performance for a typical experiment is shown in Fig. 5.10(a). The normalized error function E is shown plotted against the number of learning cycles. It can be seen that 50 learning cycles are required to reduce the error to near zero (0.02%). This corresponded to a CPU time of 13 minutes for the eight patterns on a single-user Microvax II. This is a result of the high dimensionality of the problem; for example, there

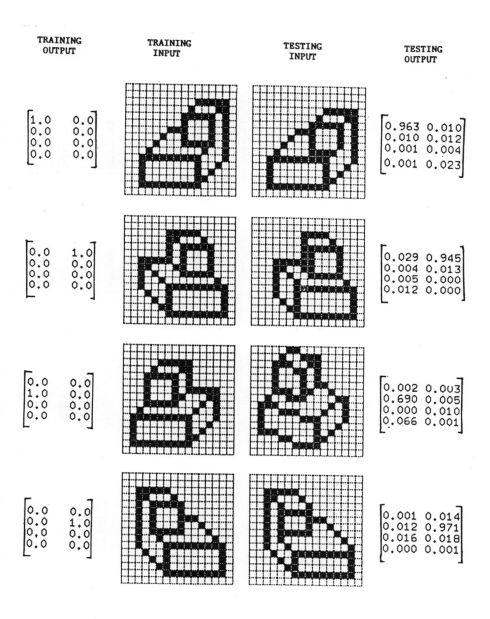

Fig. 5.7(a)   Training and test views for 16 × 16 inputs: first four views.

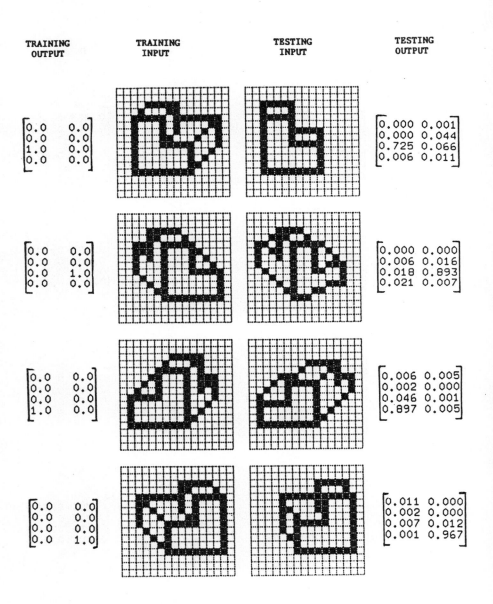

Fig. 5.7(b) Training and test views for 16 × 16 inputs: second four views.

are 4250 connections to the hidden nodes. Once trained, recognition of test views is fast, taking about one second per pattern.

Addition of noise to test data

Using the same network the experiments were repeated for a few views but now with noise added in the form of distortion to the view or the addition of random points. This is shown in Fig. 5.8.

Again the results shown that the test views and their positions are successfully recognized, now in the presence of noise. This implies that the method is robust to noise, which is important for the case of real images. We can see that the network is generalizing successfully. For example, in the second view, a complete vertical edge is missing. This is also important for the case of real vision data.

Higher-resolution input data

One of our next pieces of work is to process patterns obtained from real objects by computer vision. As a first step towards this we have repeated the experiments with an increased number of input units. The resulting network has 64×64 input nodes, 4×8 hidden nodes and 2×2 output nodes.

The results are shown in Fig. 5.9. Again the recognition accuracy is good and shows good resistance to noise in the last example. The number of learning cycles is 32 (Fig. 5.10(b)) and the CPU time is 2 hours for the 4 patterns on the same machines. The number of connections to the hidden nodes is now 130,000. The recognition time is again fast, about 30 seconds per pattern.

5.5 Comments and conclusions

Some aspects of the use of connectionist models in geometric reasoning have been discussed. A number of experiments have been described which while using synthetic data, demonstrate that views of certain objects can be recognized successfully even in the presence of noise.

We recognize that these results are only a small part of the more general problems in geometric reasoning but they are necessary if the method is to be extended. We are now investigating the recognition of other objects, questions such as displacement, the use of data from computer vision and higher-order processing.

Geometric Reasoning

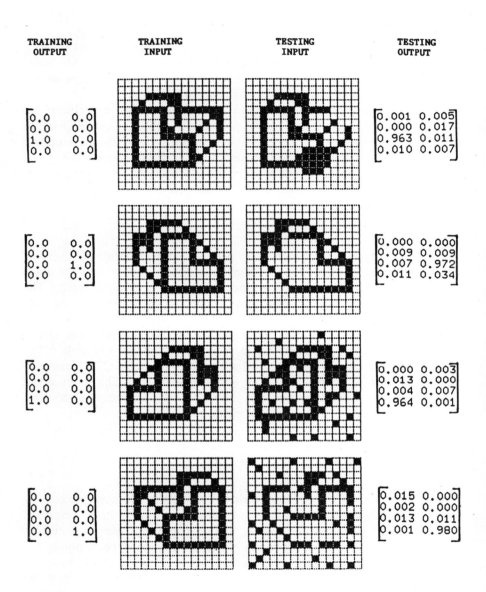

Fig. 5.8 Training and test views with noise.

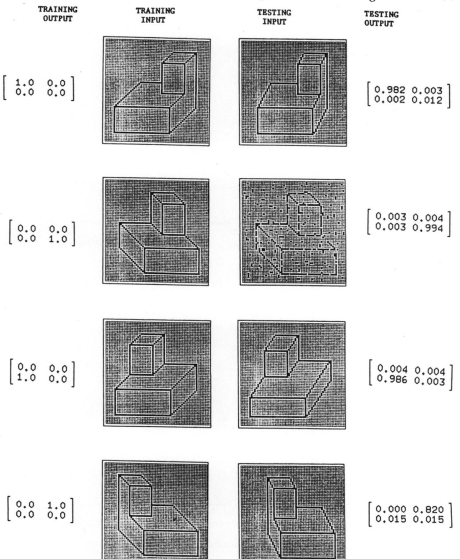

Fig. 5.9 Training and test views for 64 × 64 inputs.

5.6 Acknowledgements

One of the authors (L.-W. Chan) is supported by a grant from Kings College, Cambridge.

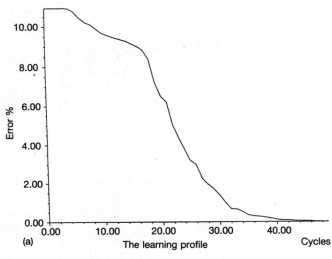

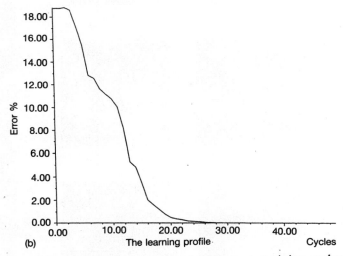

Fig. 5.10 Normalized error function E versus training cycles.

5.7 References

1. D.H. Ballard 'Cortical connections and parallel processing: structure and function', *Journal of the Behavioural and Brain Sciences* **9** (67-120), 1986.

2. A. Bundy (ed.), *Artificial Intelligence: an Introductory Course*, Edinburgh University Press, 1980.

3. P.R. Cohen and E.A. Feigenbaum (eds.), *The Handbook of Artificial Intelligence*, William Kaufmann/Pitman, 1982.

4. F. Fallside, E. Appleton, R.J. Richards, R.C.B. Speed and S. Wright 'Planning automatic assembly from a CAD database', in *Geometric Reasoning* (Proceedings of a conference held at the IBM UK Scientific Centre, December 1986) OUP, 1989.

5. M.L. Minsky and S. Papert, *Perceptrons*, MIT Press, Cambridge, Mass., 1969.

6. R.W. Prager, T.D. Harrison and F. Fallside, 'Boltzmann machines for speech recognition,' *Journal of Computer Speech and Language* 1,1 (3-27), March 1986.

7. D.E. Rumelhart and J.L. McClelland (eds.), *Parallel Distributed Processing: Explorations in the Microstructure of Cognition*, Bradford/MIT Press, 1986.

8. D.E. Rumelhart, G.E. Hinton and R.J. Williams, 'Learning internal representations by error propagation', University of California at San Diego ICS Report 8506, 1985.

9. T.J. Sejnowski and C.R. Rosenberg 'NETalk: a parallel network that learns to read aloud', John Hopkins University Technical Report JHU/EECS-86/01, 1986.

10. A.S. Tabandeh and F. Fallside 'AI techniques and concepts for the integration of robot vision and three-dimensional solid modellers', Proceedings of an International Conference on Intelligent Autonomous Systems, Amsterdam, December 1986.

5.8 Discussion

(This paper was presented by both Fallside and Chan.)

Parslow: The image you're using is *gestalt*, is it? You don't actually break it into bits and put it back together again afterwards?

Chan: At this moment we don't, because the resolution is only 64 × 64: not high enough to break it into separate objects. But, if we consider a real image, the number of pixels in each dimension will be quite high. In that case, we need to break the image into several pieces, recognize each piece, and then consider the relation between each particular component, and bring them together, and get the whole result.

Parslow: If I recall Hinton [12], he says the problem is, if you're trying to recognize a lot of objects, this technique just runs out of steam. It couldn't drive a car; it'd take you a week to work out what was coming along. You have to do it by some primitive recognition method, then you use attributes to get the right answer.

Chan: Yes. Another example that I've used is to try to make a network recognize several lines. I found it possible to train a network to recognize vertical and horizontal lines independently. If I simply add them together, it fails to recognize them. But if I modify each weight slightly, and we don't allow a choice in interpretation, it successfully recognizes two lines at the same time. I tried three lines, and that works too. Which means that if I suppress the interpretation characteristic with the excitation characteristic, it works. In other words, if I have two features present, instead of recognizing only one at a time, it recognizes that there are two objects present.

Knapman: There has been a lot of research in machine vision, rather than working purely with raw images: first of all detecting edges and then doing work on specularity, shape from shading, and lots of other things. One wonders what the point is in training a system to do edge detection when there are quite adequate programs already available, such as the Canny edge detector [11]. Why not work on the output from these reasonably well-established processes, and concentrate on training for actual object recognition at a much higher level, where there clearly is some need for showing a system a set of objects and trying to get it to recognize something, rather than trying to cope with the problems inherent in raw images, such as variations in lighting and the rest?

Chan: People have done some experiments to detect edges. They have not been all that successful, and some of the methods that people have used have been to find edges and to join them together: then to find whether some edges correspond to a particular shape. The experiment that we have done is not concentrating on edge detection alone. We might get edges from someone else's image processing technique. But the quality of the edges would perhaps not be good enough, so we couldn't rely on them; so closures would sometimes not really be closures, but discontinuous. So we try to get more from the original pixel information, by putting it into the network and detecting in that way what the correct object is.

Fallside: If I could just add to that, I think your point is quite right. In a way there is no need to do low-level processing with a network, if we have other methods of doing this, which clearly we have. What people have done, is to look at various levels of processing with these networks in an exploratory sort of way. It would certainly seem to be appropriate that, if you have good low-level processing, you can use the networks for things like transformation of processed images. It's also interesting, I think, to look at the robustness against noise. But I think your view is perfectly reasonable, particularly at present, when the efficiency of a network is so very much less than a good conventional algorithm.

Fisher: I know that this is somewhat counter to the connectionist philosophy, but did you look at the connections and the connection weights in the hidden units, to identify what features the process is using to form its classification? In some sense, your network embeds some knowledge of the object; have you done any retrospective analysis to try to find out what this knowledge was?

Chan: Yes, I've done that, but I didn't show it in the slide. I did look at the network that I used when I trained different lines. About a quarter of the hidden units are activated when I present one line: another quarter when I present the image of the second line. In effect, each hidden unit can detect one or other of the lines in the example that I tried.

Attwood: It almost looked from your diagram as though, for the 64 × 64 size image, adding noise actually improved the accuracy of recognition.

Dodsworth: Isn't it just the same as dithering in computer graphics; that adds noise but helps the human perception of pictures? Isn't it just the same principle? It's counter-intuitive, but it's possible.

Fallside: In terms of arithmetic, it's probably better to have a bit of noise, in particular where you've got these very idealized features. Other networks such as Boltzmann machines flourish in the presence of noise; they are noisy algorithms themselves, so this effect is not too surprising—at that level of noise.

M. Williams: One slightly surprising thing: looking at the top and the bottom example (in Fig. 5.9), which represent approximately the same amount of turning in one direction and then in the other, we see a 2% change in result for the top model and about 20% change for the bottom. Have you any explanation for that?

Chan: The only explanation is that the learning rules are set up internally, so I can't tell it to learn one harder than the other. In this case maybe it tries harder to learn this one than that one: so it can remember one kind of image better than the other.

Martin: There is a slight asymmetry of the position of the object in the bottom one; it may be the translational shift which is causing the difference.
  Can you say how some of these techniques relate to purely statistical techniques such as discriminant analysis, where there is no backwards and forwards iteration, but more of a direct computation?

Fallside: One point is that these are non-linear functions, and in this case there is no precise solution; you have effectively to do a small signal analysis. Some pattern analysis techniques are not non-linear, and therefore you have a straightforward solution. People are beginning to look at links between conventional pattern analysis and these machines.

Prior: Do you have only one training run, or do you give it a variety of images to train on?

Chan: If you want you can have more variety of images, but the problem is that this takes time; so I use one view for each position, instead of several views of similar positions.

Fallside: It's trained to deal with all views of the same object, so that these are put in and all that you change is the target output. It will train itself to a particular set of weights, which then remain constant when you want to recognize. If you apply any of the four right-hand test inputs, the network is now fixed. Each different view will bring up a different output. This is the recognition part: to detect what the object is by looking at the output, after training.

Prior: You never saw any slight variation on that?

Fallside: In this case it didn't; but it would if you put a whole host of class data in. As my colleague said, she had only a limited amount of time. Normally, you would put as much data in as you had available.

Chan: If you vary the orientation of the object slightly, the main features remain the same, so we try to detect the features.

Middleditch: How far can you rotate those objects and still get good results, like the ones you present? Do the results go bad when the connectivity of the image changes? Is that in fact the point when they go bad?

Chan: The figures show the most extreme case that I have tried; actually, the range contains a special case, but the system can still do the job.

Middleditch: That's the limiting case of changing the connectivity : what happens when you flip? Is it still only perturbing the output a little bit? Suppose you go a little further than that.

Chan: I think it will try to change the whole output by a sudden jump; but I have not tried it yet.

Additional references

11. J. Canny, 'Finding edges and lines in images', AI Memo 720, Massachusetts Institute of Technology, 1983.
12. G.E. Hinton, 'Some demonstrations of the effects of structured descriptions of imagery', *Cognitive Science* **3** (231-250), 1979.

6 Recognition and generation of symbolic diagrams

A. D. GEISOW

6.1 Introduction

From the earliest days of computing, there has been a desire to make computers produce and understand pictures. Until the advent of cheap memory, bit-mapped displays and thence personal computers with sophisticated graphics capabilities, graphics tended to be a specialist activity reserved for those who could justify the cost of special terminals. Now there is an explosion of interest in the application of graphics, which reflects the changing expectations of users from: 'This machine can process information for me' to 'This tool can help me solve my problems'. The latter requires much richer communication between man and machine; and as human problem-solvers typically use diagrams with which both to think and to communicate, there is naturally a desire to have the computer communicate in terms of diagrams as well.

A major theme of this paper is the use of graphics for input as well as output. Text is served by a vast array of editors which can be used to prepare input for a whole range of applications, compilers, etc., while graphics editors are either highly specialized, or are general but feed no applications at all. Those systems which do allow some form of graphical input are highly specialized in the sense that they allow the user to interact with a data-structure which is oriented towards the application, and which can be displayed graphically as a side-effect. This means that the 'graphical interface' cannot be divorced from the application, as the graphics code must have an intimate knowledge of the application's data. Generalizing an interface to facilitate use by a wide variety of systems would mean designing the data-structure around the graphics, allowing the creation of arbitrary drawings. These might then constitute illegal or meaningless input for any particular application, so the challenge is to allow this degree of flexibility while being able to check or constrain the syntax and semantics of a diagram when required. This is not to decry specialized graphics interfaces,

An example – qualitative reasoning

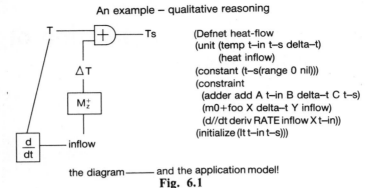

(Defnet heat-flow
  (unit (temp t–in t–s delta–t)
    (heat inflow)
  (constant (t–s(range 0 nil)))
  (constraint
    (adder add A t–in B delta–t C t–s)
    (m0+foo X delta–t Y inflow)
    (d//dt deriv RATE inflow X t–in))
  (initialize (lt t–in t–s)))

the diagram———— and the application model!

Fig. 6.1

which will always have the potential to outperform general ones; but specialization restricts re-use both of the software and of the user's knowledge.

We finish this introduction with an example. Fig. 6.1 is derived from work on qualitative reasoning [11], but for the purposes of this paper it is just an example of the use of computers to aid human problem-solving. The diagram on the left represents an idealized model of the qualitative relationships involved when heat flows into an isolated chamber of gas and the resulting temperature changes. The LISP-like notation alongside is the representation required by the machine to perform the analysis. Apart from some initialization which has been added, there is a direct equivalence between the diagram and the text. Given the preference of the user to work with the diagram, and the hard work involved in actually providing him with it, the provision of tools analogous to compiler-compilers or state-transition-driven interpreters to support the creation of this kind of graphical interface seems overdue.

6.2 Motivation

Context

Fig. 6.2 attempts to classify three application areas with respect to domain generality and type of representation. Presentation graphics (e.g. charts, graphs, business diagrams) are concerned with the display of symbolic information, and pictures are usually derived directly from that information. Systems which address this area tend to be very specific in the sense that they offer a fixed, limited range of types of picture such as bar charts, pie charts or graphs. At the other end of the spectrum, image-handling systems can display almost any kind of graphics, but work with a very direct, analogical representation. In

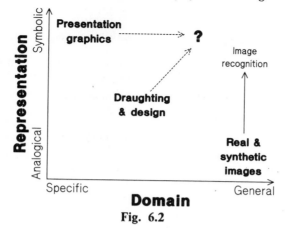

Fig. 6.2

between, there are applications such as engineering drawing, which are served by tools using mixed representation: partially symbolic (such as icons, text, lines stored as endpoints, and components represented by their shape). Such 'draughting' systems can produce a wide class of pictures, but are not as general as systems which manipulate images. Image recognition can be seen as an attempt to derive symbolic information from images: a move up the 'representation' axis of Fig. 6.2. In practice, much effort has been expended coping with noise and quantization artefacts, before structural analysis can take place. Fig. 6.2 also shows some alternative routes towards the top-right corner: either generalizing the range of presentation graphics, or moving from draughting and design systems. This paper was partially motivated by such alternatives to the more usual image-recognition work.

Why graphics?

Diagrams constitute a much richer, more expressive domain than text. Description of the two-dimensional relationships inherent in pictures is at best poorly achieved by most natural and programming languages, which are highly linear in their structure [10]. The main relationships and properties of interest are:

- Symbology: this subsumes text (as serial symbol streams), so we acquire all the capabilities of text, but we can be free to create new symbols, and are not limited to serial combination.

- Juxtaposition: a two-dimensional space allows an entity to have an arbitrary number of neighbours, text only two. Both proximity and direction of juxtaposition may be significant.

- Compactness: a large amount of complex information can be presented on a diagram without overloading the viewer.

- Topology: both connectivity and nesting of entities can be conveyed extremely well with graphics; indentation of structured programs is an attempt to enhance the presentation of nesting by semi-graphical means.

- Geometry: the inter-relationship of variables can be conveyed by conventional graphs; properties such as tangency, perpendicularity, or length may all be meaningful in a diagram.

Barriers

What stands in the way of the creation of diagrammatic interfaces, if they are so desirable? Some possible obstacles are:

- Mismatch of levels.
- Imbalance between input and output.
- Lack of rules of manipulation for diagrams.

Mismatch of levels

This occurs when an application builder wishes to use a graphics subsystem to help manipulate diagrams. The diagram may most succinctly be described in terms of a set of symbols, types of line, and perhaps areas, and various constraints describing the allowable configurations of these. Standard graphics packages operate at a lower level; they will certainly provide the basic line and text primitives, and may have the ability to define new primitive symbols and some hierarchical segment structuring, but there is little or no assistance with the maintenance of constraints and relationship between entities.

Imbalance between input and output

The poor application-builder mentioned above extends the application data structures to store all sorts of graphics data and then writes routines to display a picture from this information. Editing of the picture is a side-effect of changing the application data. The graphics package is mainly supporting output not input. This is a wasteful approach: it is difficult to re-use any of the graphics; diagrams from different applications cannot be combined in other applications (e.g. document preparation) and then extracted again, and the user interface to each application is almost certainly different.

Lack of manipulative rules

Mathematics acquires much of its power from the ability to abstract away from a situation and to manipulate a symbolic representation according to a fixed set of rules (e.g. substitution or cancellation). Such manipulative ability is much rarer for diagrammatic notations. The potential of combining such an ability with the intuitive, expressive nature of diagrams is virtually untapped. One system which does appear to have attempted this combination is GRACE/CS [7]. This is a diagrammatic notation for designing concurrent systems; it comes with design operators which are intended to allow analysis, evaluation and modification of designs while preserving the 'correctness' of the design.

Potential benefits

There are three main areas of benefit that it would be desirable to achieve simultaneously:

- A common style of interface for generic operations such as creation, deletion or transformation of elements, or querying or setting of attributes.

- Assistance with layout: this requires an understanding of the structure of the diagram in terms such as connectivity, alignment and juxtaposition.

- Link to an application: being able to deduce the relationship between the structure of the diagram and what the diagram is supposed to represent to the application.

Individually, these are relatively easy to provide. It is their combination that is of interest, as we would then have a generic interface to a variety of applications and one which could provide some intelligent assistance to the user.

We can characterize one of the major application areas as design: whether mechanical, electrical, software, or other engineering. Design is inherently an interactive process of proposal and criticism or testing of some idea. The use of diagrams as a design medium is almost ubiquitous: in software engineering there are Jackson, HIPO, SADT entity-relationship, state-transition, MASCOT and dataflow diagrams; other disciplines are just as richly populated. The ability to input an appropriate diagram, to be assisted in laying it out, to use if for analysis, to have modified versions displayed are probably universal desires for designers.

Business or presentation graphics is another areas of interest. Currently, they are used exclusively to display output—to make columns of figures more digestible to the human consumer. Could anything be gained by being able to reverse this flow? Would 'what-if?' analyses be more appreciable if one could drag the bar of a bar chart up and down and watch a graph respond, instead of modifying figures in a spreadsheet? There could certainly be improvements on the output side, if not on input. Charting and graphing systems usually have a small, fixed number of types of display available. A more flexible mapping from data to diagram would be useful.

6.3   Previous work

Overview

We now review (not comprehensively) some related work. There are three broad categories of work of interest: grammatical/linguistic, functional/logical, and procedural approaches; but there are regions of overlap between these. This paper will discuss only the first two approaches; the procedural work of interest is mainly within the object-oriented paradigm which has already received much attention for interface construction.

Grammatical/linguistic approaches

These are usually represented as being methods of describing the morphology—the form and structure—of utterances in some language. A large number of papers have been published on extensions of grammatical techniques to cover pictures. These span a range from direct techniques, where the final result of rewriting is the picture, to indirect techniques, where the result describes the picture in some way.

The following is an attempt to abstract a general notion of **grammatical description** to serve as a context for the particular methods to be described. The underlying principle of grammatical techniques is that of iteratively rewriting sub-components of some object according to a set of rules. The rules have a left-hand and a right-hand side, each of which consists of pattern of entities chosen from some predefined, (usually) finite set. For generation of utterance, one starts with some distinguished initial pattern and attempts to match sub-patterns to the left-hand sides of rules. On finding a match, the sub-pattern is replaced by the right-hand side of the rule, and the process is repeated. The entities are usually divided into terminals and non-terminals, where the non-terminals are regarded as placeholders for further development, and the process ceases when there are no non-terminals remaining. Recognition proceeds in the opposite direction: pattern matching the right-hand side to some asserted utterance and replacing matches by the appropriate left-hand side. Recognition is achieved if the initial pattern is reached and a structural description is often built by recording a trace of the rules that succeed at each stage.

For conventional **string grammars,** there is just the implicit relationship of concatenation between entities (in this case character or token symbols). Pictures have a much richer set of primitive entities and more relationships and the patterns on each side of a rule need to make those relationships explicit.

Many of the grammars mentioned below are covered in books on pattern recognition [e.g. 6]. The reader is referred to these for full details and references.

Sterling grammars can be used for describing diagrams: for example by generating statements in some programming language, including calls to graphics routines. This would be a highly indirect method and Sterling grammars have been mainly used in circumstances where there exists a simple linear encoding of the shapes of interest. The classical example is chromosome recognition based on a simple, qualitative boundary description.

Array grammars form a very direct method for raster displays. They are a straightforward generalization of string grammars to arrays of symbols in two (or more) dimensions, with up-down as well as left-right concatenation relationships. They can be used to define local operators for digital image processing.

Picture description grammars is a term which has been applied to two formalisms. The work of Shaw [16] is concerned with string grammars which generate expressions of operators and operands. The operands represent directed graphic primitives with a head and tail point; the operands control the way these are to be put together (such as head to head, tail to tail, etc.). The relationships between primitives are thus somewhat limited. Others [4,12] have proposed general descriptive methods where the rules have a single non-terminal on their left-hand sides, and right-hand sides consisting of a combination of terminals and non-terminals, with the relations between them expressed as predicates. This is very close to definite clause logic formulation and more recent work in this direction is discussed in the next section.

Graph grammars form a very general class about which much has been published [13]. Graphs with labelled edges and nodes can be used to represent the structure of pictures by modelling each primitive entity (sub-picture) by a suitably labelled node, and relations between then by arcs. Relations between complexes of entities require generalization to recursive graphs, where a node may be simple or representative of a subgraph; and n-ary relations may be handled by hypergraphs, whose arcs join two or more nodes. Much of the work on graph grammars has been devoted to investigating ways of specifying the embedding of a sub-graph within a graph when a rewrite step is taken.

Web grammars are graph grammars with labelled nodes but unlabelled arcs.

Tree grammars can either be regarded as restricted graph grammars or an interpretation of string grammars which generate bracketed strings.

Plex grammars can be regarded as a variety of graph grammar and fit well with applications such as circuit diagrams. The primitives of a plex grammar are NAPEs: N-attaching point entities. Each attaching point of a NAPE has a unique identifier and productions have lists of joints specifying how the attaching points of component NAPEs are to be connected and a specification of which joints form tie-points: the attaching points of the result of the production.

Shape grammars [5] are a direct class of methods. The terminal entities are shape elements such as line segments and non-terminals are a set of oriented markers. Pattern matching of the left-hand and right-hand sides of the production rules to a shape takes account or arbitrary affine transformations.

Grammars with coordinates have terminals and non-terminals which each have a number of points associated with them [1]. Rewrite rules are augmented with functions which compute coordinates of the replacement symbols from the coordinates of the replaced symbols. These can be regarded as a special case of the more general picture-description grammars mentioned above.

Functional/logical approaches

Whereas the work of the previous section has been largely concerned with ways of distinguishing the form and structure of a class of pictures—a 'language'—from pictures in general, another approach is to start from methods for describing the structure of general pictures. This is a matter of emphasis: whereas the former asks, for example, 'What distinguishes the class of circuit diagrams from other diagrams?', the latter asks 'How can one represent pictures such as circuit diagrams?' Simple graphical display lists, structured display files and procedural representations fall into this category but we will concentrate on declarative work in this section.

An example of the use of a highly structured description is given by Arya [2]. The applicative language HOPE is used to produce tree representations in terms of simple outline primitives and constructor functions such as above, besides and over. These representations are flattened to one level before being displayed.

A relational approach is taken in the Winchester Graphics System [9], which stores application data in a very flexible relational database. Graphics (mainly point and line) can be produced from relations is they contain appropriate coordinate and attribute fields. Required views of the application data are obtained by creating suitable relational expressions.

The objective of being able to give '...logical descriptions of the *structural mapping* between objects and their graphical representations that can be used both for graphical output and graphical input' is presented by Pereira [15]. His method is close to the relational representation but with the extra power afforded by definite clause logic. A similar motivation and approach has been reported by Helm and Marriot [8]. They present a declarative picture specification language for which they implemented an interpreter in MU-Prolog. The specifications are hierarchical and consist of rules defining a composition of sub-pictures together with relations between attributes of those sub-pictures. Logic programming, in the form of Prolog, has one of its roots in grammatical formalisms and this approach bears a strong relationship to the picture description grammars mentioned previously.

A final form of declarative representation close to the relational/logical forms is constraint-based graphics of which Juno [14] is a recent example. Juno is based on a version of Dijkstra's calculus of guarded commands. The only primitive is the point, lines and regions being in terms of these by Juno commands. Such systems rely on numerical equation solvers rather than relational or logical interference.

6.4 New directions

An outline architecture

We would like to provide tools to aid the prototyping and building of expressive, diagrammatic interfaces and to investigate their use for a variety of applications. We wish to do this in a general context, to enable re-use of software. For this reason, and because people seem to need to work at a number of different levels of abstraction in order to be efficient, we propose that application interfaces should look something like Fig. 6.3. The primitive graphics level directly represents the screen image in terms of entities such as symbols (0D), lines (1D) and regions (2D). This level should be general enough to cover a wide

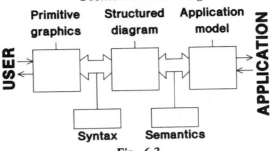

Fig. 6.3

variety of applications. For a particular application, there will be a class of 'correct' diagrams for which one writes a syntax in some appropriate grammar. This can be used to drive the translation between the primitive level and an explicitly structured one. Following this is a semantic mapping between the diagrammatic structures and those pertinent to the application.

In use, interaction would be directly with the low-level graphics. These are intended to comprise a more structured object—a diagram—which in turn represents a model driving the application. If the primitives are made to conform to the syntax, then the user is provided with intelligent editing of the diagram; if not, then unconstrained drawing or annotation would be allowed. This choice of modes gives us the flexibility to allow users to reach legal states via a sequence of illegal ones, which may be easier or more natural than being forced through 'correct' translations. As the primitives are generic to a number of applications, manipulation at this level would be common to all, increasing consistency of interface. Given an explicit representation of the syntax of the diagram, we can partition this into an essential, structural component, and a non-essential, layout component. Allowing the user to edit the latter would increase control over presentation.

If the formalism for expressing the syntax is suitably modular, simple diagram types can be re-used and combined into more complex ones. This would aid system building and contribute to more consistent interfaces. The application can work on an appropriate model without having to look after geometric information which is not of relevance to it, and present information to the user by pushing back through the two mappings.

Relation to previous work

Comparing this approach with those outlined in our description of previous work, we see that the grammatical work concentrates on the syntactic mapping component, whereas the functional/logical approach emphasizes the structured representation of the diagram. We would like to see a balanced architecture, probably using formalisms akin to either picture description or graph grammars for the mappings; and relational or logical representations for the primitive graphics, structured diagram and application models. These choices enable both recognition and generation to take place, and let the user work at a number of levels of abstraction, as appropriate. This would give a very powerful framework for creating the sort of support tools for graphics that are currently available to the designers

The requirements to support such an approach include a concise and readable notation for the syntactic and semantic mappings, and a modular environment for the relational/logical representations. Whilst simple examples can readily be expressed in variants of the formalisms previously mentioned, more realistic, detailed examples rapidly become verbose. One naturally desires a graphical notation, then machine support for it, and so one comes full circle in what is a classic example of bootstrapping.

6.5 Conclusions

We have proposed an architecture for supporting graphical interfaces based on the notation of recognition and generation of symbolic diagrams. Some of the elements of such an architecture have been considered by previous authors, but there are some challenges in the combination and context presented here. The design of usable formalisms for the syntactic and semantic mapping is probably the biggest challenge.

Although we have concentrated on *symbolic* diagrams, results in this area could filter down and to the right of Fig. 6.2, contributing to more analogical applications. Staying at the symbolic end, many people have expressed a desire for two-dimensional languages (eg PLAN2D [3]). A stronger, more flexible formal basis for diagrams would encourage more experiments, and hopefully some successes.

References

1. R. Anderson, 'Syntax-directed recognition of handprinted two-dimensional mathematics', in *Interactive Systems for Experimental Applied Mathematics* (M. Klerer and J. Reinfelds, eds.) Academic Press, New York, 1968.

2. K. Arya, 'A functional approach to picture manipulation', *Computer Graphics Forum* 3 (35-46), 1984.

3. E. Denert, R. Franck and W. Streng, 'PLAN2D—towards a two-dimensional programming language', *GI-4 Jahresstagung*, Lecture Notes in Computer Science **26**, Springer, 1975.

4. T. Evans, 'Grammatical inference techniques in pattern analysis', in *Software Engineering 2* (J. Tou, ed.), Academic Press, New York, 1971.

5. J. Gips, *Shape Grammars and their Uses*, ISR **10**, Birkauser, 1975.

6. R.C. Gonzales and M.G. Thomason, *Syntactic Pattern Recognition*, Addison-Wesley, Reading, Mass., 1978.

7. M. Harada and T.L. Kunii, 'A recursive graph theory', Proceedings of an IEEE Computer Society Workshop on Visual Languages, 1984.

8. R. Helm and K. Marriot, 'Declarative graphics', Proccedings of the Third International Conference on Logic Programming (E. Shapiro, ed.), Lecture Notes in Computer Science **225**, Springer Verlag, Berlin, 1986.

9. T.R. Heywood, B.N. Galton, J. Gillet, A.J. Morffew, P. Quarendon, S.J.P. Todd and W.V. Wright, 'The Winchester graphics system, a technical overview', *Computer Graphics Forum* 3 (35-46), 1984.

10. R.R. and M.A. Korfhage, 'The nature of visual languages', Proceedings of an IEEE Computer Society Workshop on Visual Languages, 1984.

11. B. Kuipers, 'Commonsense reasoning about causality: deriving behaviour from structure', *Artificial Intelligence* **23** (169-203), 1984.

12. R. Mohr and G. Masini, 'Drawing analysis and computer aided design', in *Artificial Intelligence and Pattern Recognition in Computer Aided Design* (Latombe, ed.), North-Holland, Amsterdam, 1978.

13. M. Nagl, 'A tutorial and bibliographical survey on graph grammars', *Graph Grammars and Applications to Computer Science and Biology* (G. Goos and J. Hartmanis, eds.), Lecture Notes in Computer Science **73**, Springer Verlag, Berlin, 1979.

14. G. Nelson, 'Juno, a constraint-based graphics system', *ACM Computer Graphics* **19**,3 (235-243), 1985.

15. F.C.N. Pereira, 'Can drawing be liberated from the von Neumann style?', in *Logic Programming and its Applications*, (M. van Caneghem and D.H.D. Warren, eds.), Ablex, 1986.

16. A.C. Shaw, 'A formal picture description scheme as a basis for picture processing systems', *Information and Control* 14 (9-52), 1969.

Discussion

Crapo: What goes on at the next higher semantic level?

Geisow: I haven't looked at that in as much detail. I've concentrated on the component that doesn't seem to be available at the moment. Once you've got a structured description, then that's beginning to get quite close to what you want

Crapo: It's a question of evaluation of symbols as structures.

Geisow: That's right. You're extracting structure from a fairly rich description, which still has embedded in it notions of geometry. You're beginning to throw away information; it's reasonably straightforward. It's when you try to build things up that it's a bit more tricky.

Fisher: Do you see a role for some sort of hand-editing phase afterwards? A certain amount of diagram might be compiled from your description, and then you might want to move some things around a little bit, while maintaining the existing geometrical constraints.

Geisow: That would be the nice thing about this sort of architecture, if one could build it. You can either talk at a dumb level, or you might choose to talk at a level where it understands the structure. It has a description of the class of diagrams, and so you can talk at those different levels of structure. If you want to drag things around and have connectivity maintained, then you should be able to achieve that simply by perturbing the thing at the appropriate level.

Martin: Systems of this kind exist in a way, but for text rather than pictures. For instance, the E-MAX editor is extensible in the sense that you can write bolt-on bits of syntax yourself for dealing with particular things, so that the editor understands in a limited way the constructs of a particular programming language or other text that you input. So, in a sense, it can take your primitive input, and turn it into a more structured thing. To push the analogy even further, you can then ask it to invoke the compiler on what it's produced, so it does the next step as well.

Geisow: Yes, but as you say, only for text, and also only processing in one direction: it tends not to work the other way around! And I think for E-MAX the understanding of the structures is fairly simple. Working even at that level with pictures would be a considerable advance on what seems to be available at the moment.

Todd: The TELL system at IBM San Jose [17] has quite a lot of what you want in it. It doesn't have by any means all, and I don't know the system well enough to be able to answer detailed questions on it, but it has got a lot of the inside knowledge of the graphic structure, so you can edit it in that sort of way.

Geisow: There are systems that have components of what I want; I'm interested in putting those components together. Human problem solvers sit down and work with diagrams, but they work with their problem at a huge number of different levels of abstraction.

Todd: I think that's the sort of thing TELL was trying to do.

Geisow: Constraint-based systems are doing something similar. There you're embedding a structure, and then playing around with it. In a sense you're doing a little bit of programming, and you're not able so much to work at any other level. It certainly hasn't got all the things one wants. It works on a particular structure, rather than a class of structures, which is fine for the one-off user, but it would be nice to be able to deal with fairly complex sort of diagrams such as circuits and other engineering graphics; and, if you're dealing with a class, you really want an application-builder to come in and be able to say what the general constraints are, to save them having to be built up from scratch each time.

Additional reference

17.   D. Weller and R. Williams, 'Graphic and relational database support for problem solving', *Computer Graphics* **10**,2 (183-189), 1976.

7 Sketch form data input for engineering component definition

C. SUFFELL and G. N. BLOUNT

7.1 Introduction

Computer software packages for three-dimensional wire-frame and surface patch component description have been available commercially for a number of years. Conventionally, the data input to these packages tends to originate from the digitization of either a three-dimensional component model, or the combination of digitized data from two or more orthographic views of the component.

Component design and development often takes the form of artist's sketches. These are translated into a physical model from which the computer description is digitized for use in subsequent design and development activities. However, preliminary studies of the component geometry at a stage prior to the development of the physical mock-up may be desirable in an effort to reduce the design and development lead-times.

It is apparent that a number of different types of sketch data may be available at an early stage of component design. These range in complexity from simple orthographic sketches, where a number of discrete views are required adequately to define the object, to pictorial sketches and, finally, more visually-realistic perspective sketches. Any combination of these sketch types can be used to provide a basic component description. Indeed it is the practice in the motor industry for clay modellers to use a range of stylist's sketches as a source of information for the generation of extremely realistic models of complete full-size models of motor vehicle bodies (and indeed vehicle interiors etc.). The clay models are then typically used for styling appraisal and ultimately, in conjunction with computer-controlled three-dimensional digitizers, as the source of data for a full three-dimensional computer model.

It would appear that an opportunity exists to circumvent the development of the physical model and to move directly from the stylist's sketches to a computer model. However, it becomes clear that sketches will inherently contain less information than is required to produce a full, comprehensive computer-based model, and additionally the reliability of the data taken from an artist's sketch, in terms of dimensional values, may be less than is ultimately expected and required.

It is hence important to identify the uses to which a computer model may be put during the design and development phase of component creation, since a model generated with the sketch information as source data is likely to be available very early in the design programme, but is not likely to have very good dimensional accuracy.

In the early design stages, general shape validation is likely to be an activity that could be catered for by a fairly coarse model where detailed information is less important than a reasonable description of the overall shape. Equally, certain geometric analyses requiring only fairly approximate data, for instance volumetric and packaging analysis, could be catered for by such a model. As the design process progresses, more precise information is required for analysis activities, for example as a means of mesh generation for finite-element analysis, or for aerodynamic characteristic investigation. In the final phase the model may be used for the definition of surfaces for machining in the production of press tools etc.

It is in the early stages of the design activity that benefit could be gained from a computer model derived from sketch data and so an investigation has been undertaken to determine whether appropriate models can be created from the types of sketches available at a sufficiently early stage of the component design programme.

7.2   Overview of the proposed solution

Computational stereo [2] has been postulated as being a convenient and efficient means of acquiring 'depth' information from a series of images of an object or scene. Indeed the process of computational stereo has been sufficiently developed as to be defined by the following steps:

- Image acquisition.
- Camera modelling.
- Feature acquisition.
- Image matching.
- Distance (depth) determination.

The above steps can be considered to be the general process flow by which computational stereo may be defined, the particular application to which the process is applied may, by its very nature, dictate that

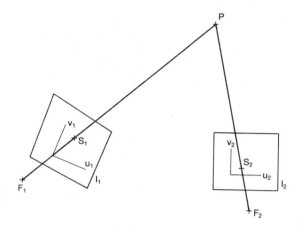

Fig. 7.1 Defining a stereo viewing system

some of the steps become trivial. In essence the process may be described graphically as in Fig. 7.1. The stereo model consists of two eye points F_1 and F_2 with their associated image plans I_1 and I_2. The rays from the object point P to the respective eye points intersect the image planes at the image points S_1 and S_2 respectively. With a knowledge of the image points and the parameters of the individual eye points and image planes, the three-dimensional location of point F may be found.

The various steps in the computational stereo paradigm described above may now be inspected and viewed in the light of the particular application to a data source from sketches.

7.3 The computational stereo paradigm applied to sketch data

Image acquisition

Various classes of sketch may be utilized as a basis for image acquisition. In the first instance a range of orthographic projections may be used (Fig. 7.2). Conventionally the orthographic views would be annotated by the addition of dimension information. In association with the inclusion of hidden detail, this allows the extraction of realistic geometric information from the combination of views.

In contrast to orthographic information, a series of pictorial sketches may be considered as a form of source image. It is conventional to include neither dimensions nor hidden detail in a pictorial sketch. However, a series of rules governing the composition of the pictorial

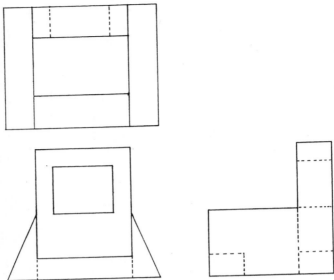

Fig. 7.2 Third angle orthographic projection.

projection need to be adhered to. Principally these rules require that any lines parallel on the object will remain parallel in the sketch and that dimensions measured in any of the three orthogonal directions on the object will be required to retain a scaled equivalence in the sketch.

The very nature of these conventions will dictate that the use of true pictorial sketches will be limited when considered for use in the design process. This is particularly true when one considers that the fundamental advantage of pictorial sketches is that they produce 'picture-like' or more specifically 'visually realistic' images. However, they offer only a crude approximation to the true viewed image at the cost of fairly complex rules of generation. These rules mean that pictorial views may be constructed accurately but, since we are primarily interested in sketches, they form a less favourable view type. Perspective sketches are more closely allied to the 'styling' sketches produced by the concept designer at an early stage of component development, and represent the most likely form of data presentation available to the present investigation. However it is quite possible that any of the foregoing forms of sketch may be available, and consequently a technique is required that will cater for all three forms of sketch input. As will be shown, orthographic and pictorial sketches may be considered to be no more than special cases of the perspective view of an object, hence a scheme that caters for the latter should equally cater for the two former cases.

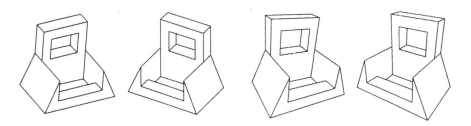

Fig. 7.3 Pictorial views
of a component.

Fig. 7.4 Perspective views
of a component.

Camera modelling

In order to allow the use of two ostensibly two-dimensional images of an object as a means of defining the three-dimensional geometry, certain information is required with regard to the projection geometry for the individual images to be used. In the case of photogrammetry, where the source of image transformation is a pair of photographs, then the projection geometry for each individual photograph consists of the information shown in Fig. 7.5. A similar geometric model may be constructed where the source of basic information is a sketch rather than a photograph. However, the definition of the projection geometry becomes more complex.

A variety of camera resection schemes have been postulated. Staffeld [5], in his work on motor vehicle crash analysis, utilized a method based on an initial estimate of the camera parameters and, by means of a least-squares error fitting technique, iteratively improved the estimated location of the camera and defined the associated camera focal length. The model involved an interactive approach to finding the camera properties and a knowledge of at least six reference points on the object. They would typically be the centres of the front and rear wheels, the centre-line section, and critical points on front and rear bumper lines: the conventional known points of a motor vehicle.

Yakimovsky and Cunningham [7], in their work on the Jet Propulsion Laboratory Robotics Research Vehicle, utilized an approach to camera modelling based on the vector representation of the camera parameters. To allow a solution to the vector equations a knowledge of eight points on the object is required.

An alternative approach is taken by Rogers and Adams [4] and by Wu et al. [6] inasmuch as the camera model is considered to consist of a suitable perspective transformation matrix which may, by dint of a

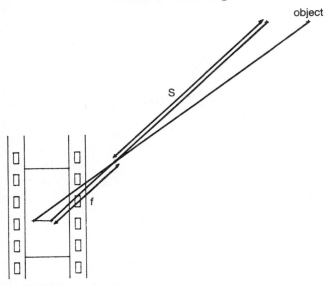

Fig. 7.5 Projection geometry for photogrammetry. **S** is the subject distance
and **f** is the approximate focal length.

process of matrix manipulation, be derived as follows, for a single
camera arrangement as described in Fig. 7.6.   If we consider a
homogeneous coordinate system (as discussed by Ballard and Brown [1])
in both 3-space and 2-space, then Equation 7.1, equating the image
points to the equivalent object points, holds:

$$
\begin{bmatrix} T_{11}T_{12}T_{13}T_{14} \\ T_{21}T_{22}T_{23}T_{24} \\ T_{31}T_{32}T_{33}T_{34} \end{bmatrix}
\begin{bmatrix} x_1 x_2 \cdots x_n \\ y_1 y_2 \cdots y_n \\ z_1 z_2 \cdots z_n \\ h_1 h_2 \cdots h_n \end{bmatrix}
=
\begin{bmatrix} U_1 U_2 \cdots U_n \\ V_1 V_2 \cdots V_n \\ f_1 \ f_2 \ \cdots \ f_n \end{bmatrix}
\tag{7.1}
$$

where the location of the nth point in 3-space is $(x_n/h_n, y_n/h_n, z_n/h_n)$, and
its equivalent projected image point is $(u_n/f_n, v_n/f_n)$.
    Equation 7.1 may be re-written as

$$\mathbf{TP} = \mathbf{S}. \tag{7.2}$$

To solve this form of matrix equation a least-squares technique may be
employed, so we express the error as

$$\mathbf{E} = \mathbf{TP} - \mathbf{S} \tag{7.3}$$

and so, squaring:

$$\mathbf{E}^2 = (\mathbf{TP} - \mathbf{S})^2. \tag{7.4}$$

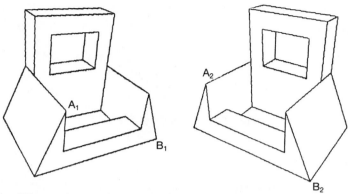

Fig. 7.6 'Camera' geometry for perspective transformation. A_1 is 'equivalent' to A_2 and B_1 is 'equivalent' to B_2 etc.

We now minimize the square of the error with respect to the elements of the matrix **T**, that is differentiate **E** with respect to **T** and set to zero. Hence,

$$(\mathbf{TPP}^T) - (\mathbf{SP}) = 0 \tag{7.5}$$

or

$$\mathbf{T} = \mathbf{SP}(\mathbf{PP}^T)^{-1}. \tag{7.6}$$

Our manipulations culminate in a matrix equation allowing the solution of the 4×3 transformation matrix **T**. The variables input to allow the numerical solution to be the measured image points (u, v, f) correspond to known reference points (x, y, z, h), on the object. The significance of this is that, with the knowledge of at least six points on the object, we may fully define the transformation matrix associate with the camera image (or in the case of this investigation the sketch).

The 4×3 transformation matrix may be partitioned in the following manner:

$$\begin{bmatrix} T_{11} T_{12} T_{13} & T_{14} \\ T_{21} T_{22} T_{23} & T_{24} \\ T_{31} T_{32} T_{33} & T_{34} \end{bmatrix}, \tag{7.7}$$

where the 3×2 partition applies a linear transformation (i.e. involving a linear combination of scaling, rotation and translation), the 1×3 row matrix produced the perspective projection, the 2×1 column matrix gives translations and the final element T_{34} contributes an overall scaling factor.

As the evaluation of Equation 7.6 represents at least-squares approximations to the transformation matrix (Equation 7.7), the error may be estimated as

$$E_c = \sum_{i=1}^{n} \sum_{j=1}^{3} \sum_{k=1}^{4} T_{jk} P_{ki} - S_{ji},$$ (7.8)

where n is the number of calibration points ($n \geq 6$).

Feature acquisition

For a variety of applications, automatic feature acquisition may be of importance but, for the current study, manual methods may well be considered to be wholly adequate. Indeed since the application of the computational stereo paradigm is envisaged as being an integral part of the conceptual design process, interaction with the designer could well be considered to be a fundamental requirement.

Image hatching

To enable the computational stereo process to function as we are required to identify 'equivalent' points on two (or more) source sketches, that is to identify the image points that have been mapped as being the view transforms from the same object point (Fig. 7.5), envisaged as being an integral part of conceptual design. This process may be satisfactorily achieved interactively, since the designer can effectively identify 'equivalent' points aided by his inherent knowledge of the component in question.

As will be seen later, the above assumption may need adaption depending upon the type of data to be acquired. However, current software development has indicated that the use of a simple tablet and stylus, in the hands of a designer familiar with the component to be modelled, can provide appropriate input data.

Depth determination

Equation 7.7 is formed for a single calibration point. If the homogeneous coordinates of the object point are normalized, i.e. $h_{i=1,n} = 1$ then:

$$T_{11}x + T_{12}y + T_{13}z + T_{14} = \frac{u_1}{f_1}$$ (7.9)

$$T_{21}x + T_{22}y + T_{23}z + T_{24} = \frac{v_1}{f_1} \tag{7.10}$$

$$T_{31}x + T_{32}y + T_{33}z + T_{34} = f_1. \tag{7.11}$$

so, using f from Equation 7.11 in Equations 7.9 and 7.10,

$$\begin{aligned} x(T_{11} - T_{31}u_1) + y(T_{12} - T_{32}u_1) \\ + z(T_{13} - T_{33}u_1) + (T_{14} - T_{34}u_1) = 0 \end{aligned} \tag{7.12}$$

$$\begin{aligned} x(T_{21} - T_{31}v_1) + y(T_{22} - T_{32}v_1) \\ + z(T_{23} - T_{33}v_1) + (T_{24} - T_{34}v_1) = 0. \end{aligned} \tag{7.13}$$

Equations 7.12 and 7.13 represent two plane equations for a point (u, v) in the image plane. The intersection of these two planes represents a line from the point on the object, through the image point. By determining the intersection of the two lines derived from two separate views of the same point we may obtain the true three-dimensional location of that point.

Thus we may write the equations equivalent to 7.12 and 7.13 for the two views of a single object point:

$$\begin{aligned} A_{11}x + A_{12}y + A_{13}z &= B_1 \\ A_{21}x + A_{22}y + A_{23}z &= B_2 \\ A_{31}x + A_{32}y + A_{33}z &= B_3 \\ A_{41}x + A_{42}y + A_{43}z &= B_4, \end{aligned} \tag{7.14}$$

where the following hold:

$$\begin{aligned} A_{11} &= Ta_{11} - Ta_{31}u_a \\ A_{12} &= Ta_{12} - Ta_{32}u_a \\ A_{13} &= Ta_{13} - Ta_{33}u_a \\ B_1 &= Ta_{34}u_a - Ta_{14} \\ A_{21} &= Ta_{21} - Ta_{31}v_a \\ A_{22} &= Ta_{22} - Ta_{32}v_a \\ A_{23} &= Ta_{23} - Ta_{33}v_a \\ B_2 &= Ta_{34}v_a - Ta_{24} \\ A_{31} &= Tb_{11} - Tb_{31}u_a \\ A_{32} &= Tb_{12} - Tb_{32}u_a \\ A_{33} &= Tb_{13} - Tb_{33}u_a \\ B_3 &= Tb_{34}u_b - Tb_{14} \\ A_{41} &= Tb_{21} - Tb_{31}v_a \\ A_{42} &= Tb_{22} - Tb_{32}v_a \\ A_{43} &= Tb_{23} - Tb_{33}v_a \\ B_4 &= Tb_{34}v_b - Tb_{24}. \end{aligned}$$

In the above relationships, u_a and v_a refer to the image point in view 1 and u_b and v_b refer to the corresponding image point in view 2. Similarly Ta_{ij} refers to an element in the transformation matrix for view 1 and

Tb_{ij} refers to the equivalent element in the transformation matrix for view 2.

Equations 7.14 may be seen to be a set of four equations in three unknowns, namely x, y, z (the three-dimensional location of the object point). Consequently a least-squares method can be employed to solve for the unknowns.

Equations 7.14 may be re-written in matrix form as:

$$AX^T = B, \qquad (7.15)$$

where the elements of **A** are described above,

$$X = [x, y, z]$$

and

$$B = [B_1, B_2, B_3, B_4].$$

Following a series of matrix manipulations we determine that

$$X = (A^T A)^{-1} A^T B. \qquad (7.16)$$

Thus with a knowledge of the transformation matrices for the two views and the coordinates in 2-space from the two views of an equivalent point in 3-space, the 3-space coordinates may be found.

7.4 The computational stereo paradigm applied to sketch input

Faceted objects

The methods for depth determination from a pair of sketches may be applied to a component formed from a series of planar facets. In this context a planar facet is considered to be flat bounded area which may lie at any orientation but must lie in a single plane. Since all the features of such an object may be described by reference to an ordered sequence of discrete points, then by acquiring those points and attributing the required ordering definition, a complete 'wire-frame' description may be generated.

As was implied earlier, the methods previously defined have been incorporated into a computer program implemented on an IBM AT micro-computer and the following test case has been processed. Table 7.1 represents the known coordinates of a number of points on the object described graphically in Fig. 7.4 compared with the same points when calculated using the computational stereo paradigm described above.

Known Coordinates			Calculated Coordinates		
x	y	z	x	y	z
0.8	-0.8	0.0	0.80797	-0.79658	0.00627
-0.3	-0.5	1.2	-0.29984	-0.49868	1.20022
0.8	0.5	0.6	0.77410	0.48367	0.59510
0.0	-0.31	0.59	0.00160	-0.31267	0.59118
-0.3	0.5	1.2	-0.31177	0.49264	1.19801
0.8	-0.5	0.6	0.80228	-0.49876	0.57865
0.0	-0.5	0.6	0.03994	-0.48046	0.60992
0.0	-0.31	1.01	0.01537	-0.30452	1.00316
0.0	0.31	1.01	0.00191	0.31087	1.00653
0.0	0.31	0.59	-0.02498	0.29487	0.59387
0.0	0.5	0.6	-0.01379	0.49212	0.60764
0.8	-0.5	0.19	0.79610	-0.49601	0.19141
0.8	0.5	0.19	0.74154	0.45860	0.20103
0.0	-0.5	1.2	-0.01393	-0.52057	1.18241
0.0	0.5	1.2	0.02983	0.51736	1.19913
-0.3	0.5	0.6	-0.34257	0.47816	0.61170

Table 7.1   Comparison of known and estimated coordinates.

Curved objects

Although some components may be considered to be composed of a series of planar facets, many do not conform to this specification and in particular may contain curved features. It is possible under certain circumstances to decompose curved features into a series of facets; indeed, many computer modelling systems do precisely that. However this is an approximation that may not be considered acceptable in all instances.

The inherent problem when considering general free-form curves in space from the point of view of data acquisition, is that with the exception of the end points it is not immediately possible to establish the 'equivalent' points in the views.   Since this is the fundamental requirement when applying the techniques described above, it would seem that free-form curves may not be handled.

However a solution may be envisaged if the form of the features acquired from the image is reconsidered. To date we have assumed that only discrete points will be acquired from the images, however we are now faced with having to extract curves in their own right.   If we consider curves C_1 and C_2 in Fig. 7.8. By digitizing the two individual curves we can see that although the end points may be matched we can

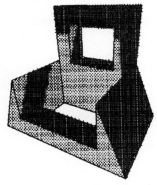

Fig. 7.7 Component defined by planar facets.

identify no other pairs of points that correspond to the same object point.

If however we assume that the 'equivalent' points lie at 'parametrically' equivalent points along each curve. That is if the curve length is parameterized and that some point, say P_1, at a distance a third of the arc length along curve C_1, is 'equivalent' to a point P_2 also at a distance a third of the arc length of curve C_2. Then the point-matching process described earlier may be used to obtain the 3-space location of the point.

To achieve this parameterization, a variety of methods are available; the one chosen here is based on the Bézier curve fitting methods [3]. It has been shown that any curve may be described by a finite number of pole points $\mathbf{B}_0 \rightarrow \mathbf{B}_n$ forming a characteristic polygon, where the following function applies:

$$\mathbf{P}(u) = \sum_{i=0}^{n} \mathbf{B}_i f_i(u) \ (0 \le u \le 1). \tag{7.17}$$

The vector $\mathbf{P}(u)$ describes the 3-space location of a point a parametric distance u along the curve. The vector $\mathbf{B}$ represents the 3-space location of the ith pole point and $f_i(u)$ is given by the Bernstein polynomial:

$$f_i(u) \frac{n!}{i!(n-1)!} u^i (1-u)^{n-i}. \tag{7.18}$$

Thus in general terms we may establish the location of any parametric point along the curve by using:

$$\mathbf{P}(u) = \sum_{i=0}^{n} \frac{n!}{i!(n-1)!} \mathbf{B}_i u^i (1-u)^{n-i}. \tag{7.19}$$

Equation 7.19 applies equally well whether the vectors $\mathbf{P}$ and $\mathbf{B}$ are in 2-space or 3-space. Hence if we may fit an appropriate characteristic

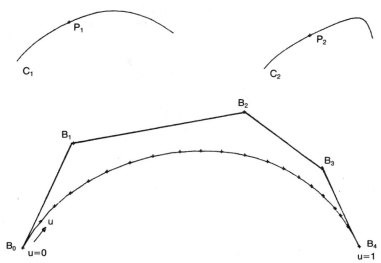

Fig. 7.8 (above) Perspective views of a free-form curve.
Fig. 7.9 (below) Characteristic polygon with five pole points.

polygon to the 2-space image curves in Fig. 7.8 then we may define the 2-space location of any point at a known parametric distance along the curve. This being the case we may now use these points as a means of generating the 3-space locations along the object curve.

It has been found that for curves of the form shown in Fig. 7.10. where the change in radius of curvature for the individual 2-space curves may be considered to be 'fairly constant', that the above procedure of approximation yields acceptable results. This is shown in Fig. 7.11. However for curves with a significant variation in the rate of change of curvature along their length, increasing error is detected in the calculated locations. An alternative approach to the feature acquisition for free-form curves is needed and an application of surfacing theory may be considered appropriate.

7.5 Conclusions

Computational stereo is seen to represent a viable process by which generally available sketches may be utilized to provide a source of data to be input to a computer based component description.

Objects composed of a collection of simple planar facets may be successfully described in this manner, but some limitations currently exist in the treatment of objects with curved surfaces. These limitations are the subject of continued investigation and preliminary proposals

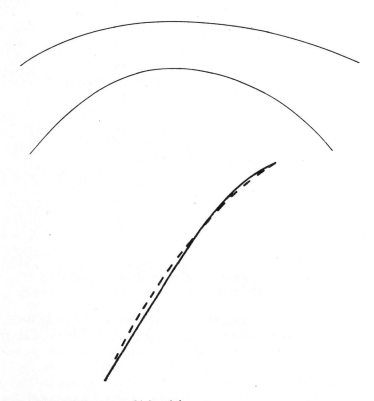

Fig. 7.10 (above) Images of 'simple' curves.
Fig. 7.11 (below) Isometric view of the original curve defined by the two curves in Fig. 7.10 (solid line) and new curve generated using computational stereo (dotted line).

have been put forward for the possible solution to the problems they pose.

The potential accuracy of the component description resulting from the application of computational stereo will in part depend on the geometric reliability of the sketches provided as source data. It seems highly likely that 'artistic licence' may be invoked in the sketches produced by the component stylist. However, the work on crash-film analysis [5] indicated that predictable errors in image formation, attributable to lens distortion may be corrected by a series of 'lens correction factors' determined by physical test. A similar approach may be considered necessary to cater for any 'predictable' distortion in sketch image, the 'sketch correction factors' being included as additional transformations into the camera model.

References

1. D.H. Ballard and C.M. Brown, *Computer Vision*, Prentice-Hall, Englewood Cliffs, NJ, 1982.
2. S.T. Barnard and M.A. Fishler, 'Computational stereo', *Computing Surveys* 14,4, 1982.
3. P. Bézier, *Numerical Control Mathematics and Applications*, Wiley, New York, 1972.
4. D.F. Rogers and J.A. Adams, *Mathematical Elements for Computer Graphics*, McGraw-Hill, New York, 1976.
5. S.E. Staffeld 'Photographic analysis of impact tests', American Society of Automotive Engineers publication 700409, 1970.
6. C.K. Wu, D.Q. Wang and R.K. Bajcsy, 'Acquiring three-dimensional spatial data of a real object', *Computer Vision, Graphics and Image Processing* 28 (126-133), 1984.
7. Y. Yakimovsky and R. Cunningham, 'A system for extracting three-dimensional measurements from a stereo pair of TV cameras', *Computer Graphics and Image Processing* 7 (195-210), 1987.

Discussion

(This paper was presented by Suffell.)

Murray: What were the major sources of error in the faceted model that you had?

Suffell: The errors were the results of two aspects of the analysis. Both during the camera modelling process and during depth determination the solution is over-defined, and so a least-squares approximation is required so as to converge on appropriate solutions. That approximation is responsible for the majority of the errors. Also, input to the scheme is currently based on physically pointing a stylus at a piece of paper. There are further errors in that process. One can only identify points to within a certain degree of accuracy. The next extension to this process will be to avoid that last form of error by utilizing a drawing scanner; I am in the middle of assembling that at the moment.

Murray: Would you ever see this technique being an input device to, say, a solid modelling system? We could actually take two stereo views of a physical object, such as a building, and create the model from them.

Suffell: That technique is used to some extent. Stereo-photogrammetry has been for a number of years one of the established methods of converting a clay model into a computer model. That relies on photographs, where one set of least-squares approximations is unnecessary, since we can identify the camera model fairly accurately. Certainly I see no reason why we couldn't use, say, architect's sketches to develop a model of a new building using this techniques, but at this stage I'm primarily interested in the use of sketches.

Martin: One thing that you don't seem to have discussed is that these are sketches and not accurate data that you're starting from. How do you cope with that? Can you also tell us how accurate styling data is likely to be?

Suffell: From the point of view of true perspective, styling data has been shown to be very accurate in terms of its proportion. Having said that, stylists are artists, and so they do like artistic license. There will be predictable distortion of images. For instance, stylists always like to put much wider wheels on a vehicle than it really has; they look smarter! In the crash analysis work at Chrysler to which I referred [5], Staffeld developed a lens correction scheme. The lenses that he was using to film the crashes weren't perfect, and there was distortion in the image. Analysis of the images taken through lenses of known patterns identified a series of lens correction factors, which basically represented nothing more than an intermediate transformation matrix in the camera model. I could envisage a similar process for sketch correction factors. His argument was that the lens correction factors were based on the fact that he could predict that each image was going to have some form of known distortion and, I suspect, having looked at a number of stylist's sketches, that similar arguments could be applied to a stylist's sketches, although I also suspect that the matrix for each stylist might be different. There are certainly possibilities along these lines.

Martin: Lens errors are surely fairly gross and vary slowly across the whole image, whereas in one sketch a stylist might draw the front door bigger than the back door, but in another the front door bigger than the back door.

Suffell: The quality of styling sketching tends to rule that out. The styling sketches are the primary and only form of input to the clay model, and so such a sketch is currently used as the means of generating the accurate and master model for the vehicle. Stylists are demonstrably sufficiently accurate to be able to produce a first-pass model. I'm not suggesting that sketches could be used to produce the final model. But they should be capable of producing a model which can support the primary geometric analyses: things like packaging and viewing parameter analysis.

Stobart: You described a system that was sequential, in that the stylist generates the sketches and then you process them. Do you think that there is any scope for an interactive technique?

Suffell: I prefaced my talk by saying that computer-aided styling—a rather horrible term—is becoming the buzz-word in the field. The ultimate aim of computer-aided styling, as I understand it, is to achieve an interactive scheme, whereby a stylist can utilize 'paintbrush software', via some form of transformation, so as to create an appropriate realistic object. That's the aim. At the moment it appears that computer manufacturers are highly delighted to work on the two ends of that spectrum: paintbrush software and visual realism. Tying the two together is not actively being followed by a large number of people. As I say, that is the ultimate aim, and current digitizing equipment technology, for instance, allows a tactile styling-type feel to data acquisition. I think the component parts of a system could be available reasonably quickly, but not now.

8 Geometric constraints from 2½D sketch data and object models

R. B. FISHER and M. J. L. ORR

8.1 Introduction

A competent vision system needs to be able to locate objects in the environment as well as to identify them. Hence, it is necessary to derive and integrate position information from data elements paired with model features. There are additional applications of geometric reasoning in the context of model-based vision:

- The prediction of image locations (or other properties) for additional evidence.

- Testing of proposed model-to-data pairings by ensuring the geometric relationships embodied in the model are upheld by the data.

The first major section of this paper summarizes a recent study [7] of several high-performance vision systems, classifying the tasks and operations used in their visual analysis. Analysis has shown that there are a few simple underlying geometric operations but that the types of models and data used drastically affect the machinery required to implement them.

Several recent vision projects here in Edinburgh [3,4] have concentrated on recognizing and locating objects using surface based models and three-dimensional surface data. There is twofold justification for this effort. First, research into recognition based on edges has shown how fragile this approach is. Second, much low-level vision research is investigating delivery of some form of three-dimensional scene description, whose advantages are:

- The interpretation of surface data is unambiguous (unlike intensity data, which can arise from several scene effects).

- Working in three dimensions provides direct information about the scene (unlike projected intensity images).

In consequence, our research has been investigating using surfaces as the primary data and model features.

To exploit the richer information in surface data, we have been investigating surface-based object representations [4]. Object primitives are represented using groups of rigidly connected surfaces, which have been segmented into regions of near-constant shape character. Larger objects are hierarchically defined by connecting previously defined sub-components, where the connections need not be rigid. Surface primitives promote direct pairing between model and data surfaces, easier extraction of object positions (using the three-dimensional data) and easier prediction of feature visibility. Related work is also investigating the role of other geometric entities in the data and models (e.g. volumetric and boundary representations).

One important research question that has arisen from the conjunction of the model and data types is: what constraints on object position arise from the pairing of a model feature with a data feature? This is complicated, because of:

- The variety of data and model features.
- The variety of constraints types obtainable from pairing such features.
- The existence of constraints that are only partial.

This paper reports our recent work in this area.

Finally, we have been investigating mechanisms for implementing the geometric reasoning functions needed for vision. Some of the difficulties in this area lie in:

- Integrating partial constraints.
- Coping with data errors.
- Finding fast mechanisms.

The last major section of this paper summarizes initial results in new research into a parallel network geometric reasoning engine. The network uses inequality constraints (the arcs) on individual geometric variables (the nodes). Integration of constraints occurs when the network converges to a consistent state. The structure of the network is predefined using some extensions of ACRONYM's [2] constraint manipulation system. This allows slow symbolic manipulation at network definition time, yet fast execution during recognition.

8.2   Geometric reasoning for model-based vision

The ability to reason about the geometry of a scene is an essential aspect of any sophisticated vision system.   Geometric reasoning can be understood at three levels:

- The various tasks that a vision system requires of a geometric reasoner.

- The ideal data types and operations required for the tasks.

- The implementations of the operations.   As we will see, it is easier to formulate the required operations than to find perfect implementations.

The following sections deal with these topics in turn.

Geometric reasoning tasks

A geometric reasoner is characterized by the tasks that it is expected to carry out.   Based on an analysis of existing vision systems, we have identified the following tasks.

Establishing position estimates

Every identified feature in an image can be used to form **position constraints,** first because the feature is visible, and second through its measurable properties (location, shape, dimensions and so on).   In order to aggregate information from related features it must be possible to *combine* individual position constraints into a single position estimate. During such combination, the detection of **inconsistent constraints** is required to eliminate false hypotheses (formed, for example, from erroneous feature identifications).

It may be necessary to transform position estimates into another reference frame before other functions are applied.   One example of this is when the position of a feature in an object's reference frame is desired, given only knowledge of the feature relative to a sub-component and the sub-component relative to the object.

If the modelled relation between parts of the same object involve degrees of freedom then there has to be some **variable binding**.

Image prediction

Given a position estimate for an object it should be possible to *predict* the appearance and location of its features This allows comparison between the predicted and observed features, and affords a basis for reasoning about occlusion effects. Additionally, image prediction can be used to search for features not already found.

Predicted features are not necessarily pixel-type entities. They may also be more symbolic entities such as points, lines, surfaces, normals and so on. Furthermore, real images are formed from objects with exact positions, but predicted images may involve objects whose positions are only roughly known, hence the predictions should be able to represent some kind of uncertainty.

Geometric reasoning functions

The second descriptive level of geometric reasoning concerns the abstract data types and operations required for the tasks described above.

Positions

Geometric reasoning requires a data type for representing positions. The traditional representation by three translational and three rotational degrees of freedom is not adequate for our purposes. Extensions are needed for:

- Uncertainty present in image measurements.
- Modelling objects with positional degrees of freedom.
- Modelling objects with size variation or tolerances.

Positions also transform points, vectors and other positions from one coordinate frame to another (e.g. the position of a sub-component is equivalent to a transformation from the object's to the sub-component's coordinate reference frame). Hence, if we extend the notion of position from a point to a region of six-dimensional parameter space, then these transformations are no longer one-to-one mappings, and we will have to deal with uncertain points and vectors.

In what follows we will be giving some simple data type specifications [5] using the operators FRAME and PLACED (capital letters will be used for all operators). Both operate on members of the set *Position* and return members of the set *Model*. The latter includes the special models *World* and *Camera* so that we can have world-centred and viewer-centred coordinate systems as well as relative positions between models. Thus we write the functionality of FRAME and PLACED as:

FRAME: *Position* → *Model*

PLACED: *Position* → *Model*.

FRAME returns the model whose frame is the reference frame of a position and PLACED yields the model placed by a position.

Estimating positions from features

Each pairing of a model to a data feature produces constraints on the position of the model to which the feature belongs. We have then an operation, LOCATE, whose inputs are the model feature and the image feature and that yields a position estimate whose FRAME is the camera and whose PLACED object is the model:

LOCATE: *Image_feature, Model_feature* $\rightarrow$ *Position* $\cup$ *undefined*
for all $f_i \in$ *Image_feature* and $f_m \in$ *Model_feature*
  let $p = \text{LOCATE}(f_i, f_m)$
  if $p = undefined$ then
    FRAME(p) = *Camera*
    and
    PLACED(p) = m,

where m is the model to which f_m belongs, $\cup$ is the set union operator, and *undefined* is a set with one member, the undefined object.

LOCATE returns *undefined* to signal an invalid pairing between incompatible image and model features.

Merging positions

In general models consist of more than just a single feature. If several features are identified in the data and produce constraints on the object position, then we need to verify geometric consistency and merge estimates.

The MERGE operation acts on sets of positions and returns a position. The result is only defined when all the input positions have the same coordinate frame and refer to the same object. The *inconsistent* result means the input positions were inconsistent. If #(*Position*) is the power set of *Position* (the set of all possible subsets of *Position*):

MERGE: #(*Position*) $\rightarrow$ *Position* $\cup$ {*undefined, inconsistent*}
for all $m_1, m_2 \in$ *model, S* $\in$ #(*Position*):
  let $p = \text{MERGE}(S)$
  (for all $q \in S$:FRAME(q) = m_1 and PLACED(q) = m_2) then
    $p = inconsistent$
    or
    FRAME(p) = m_1
  and PLACED(p) = m_2
  else
    $p = undefined.$

Transforming position constraints

Suppose we know the position of an object A relative to another object
B. This may occur if:

- They are parts of the same larger model assembly.

- We have *a priori* knowledge about their relationship (e.g. the
  position of the camera in the world).

- We observe the relationships between their features in the image
  (e.g. 'face 1 of A is against face 2 of B').

Two geometric problems then arise. First, if we know the position of
A in the frame of some other object C, what is the position of B in this
frame? Second, if instead we know the position of C in A's frame, what
is the position of C in B's frame? These problems require the operations
TRANSFORM and INVERSE that obey the following rules in relation
to the operators FRAME and PLACED:

> TRANSFORM: *Position, Position → Position* $\cup$ *{undefined}*
> INVERSE: *Position → Position*
> for all $p, q \in position$
>   let $r = $ TRANSFORM(p, q)
> if PLACED$(p) = $ FRAME(q) then:
>   FRAME$(r) = $ FRAME(p)
>   and
>   PLACED$(r) = $ PLACED(q)
> else
> $r = undefined$
> for all $p \in Position$
>   FRAME (INVERSE (p)) = PLACED (p)
>   PLACED (INVERSE (p)) = FRAME (p).

Now, if we represent by X/Y a position whose FRAME is X and whose
PLACED object is Y, our two problems can be written as:

$$C/B = TRANSFORM(C/A, A/B)$$

and

$$B/C = TRANSFORM(INVERSE(A/B), A/C).$$

Image prediction

Image prediction involves several operations that differ only in what is
being predicted. Typical image predictions are:

- Feature visibility.

- Feature orientation.
- Feature appearance.
- Feature location (image and scene).
- Feature relative depth (e.g. surface ordering and occlusion relations).

The operation to be performed in any given prediction task depends both on the task and on the nature of the feature whose image is being predicted. For example, to determine whether a plane surface is front-facing requires projecting a predicted surface normal along the line of sight, but this operation would be insufficient if the surface were curved. We abstract them all into the PREDICT operator, which must have knowledge of the different feature types:

$$\text{PREDICT: } Model\_feature, \ Position \rightarrow Pred\_feature,$$

where Pred_feature is a separate data type from Image_feature because if must incorporate variations due to uncertain positions.

Review of current implementations

Some existing AI programs, including RAPT [8], IMAGINE [3] and ACRONYM [2], have geometric reasoning capabilities. RAPT, a robot planning program, and ACRONYM, a model-based vision program, both represent and manipulate positions symbolically and exploit relationships between symbolic expressions (equalities or inequalities) for their deductive power. RAPT is less powerful as it can only deal with relationships that are equalities. In contrast to these, IMAGINE uses a numerical representation scheme with two separate position data types. One type (homogeneous matrix) is designed for the TRANSFORM operator and the other (parameter bounds) for MERGE. A problem with this scheme is that the matrix type represents exact positions whereas the parameter bounds represent rough positions and so conversion between the two forms involves loss of information or approximation.

ACRONYM represents positions using a variable for each degree of freedom. Constraints on positions are formed by relating expressions in the variables to quantities measured from the image and manipulated symbolically.

The ideal constraint manipulation system should be able to:

- Decide if each variable has a value consistent with the constraint set.
- Bound any expression in the variables over the constraint set.

The operations MERGE, TRANSFORM and PREDICT are achieved by unioning constraint sets, simplifying symbolic compositions of positions and bounding expressions in variables.

Three advantages of using an algebraic constraint representation are:

- The representation is *incremental* in two senses:

  - New constraints can be added when each new piece of evidence is discovered.
  - New classes of constraints can be added when new visual relationships are understood.

- The representation is *uniform*, because it can represent a large range of visual relationships using the same mechanism.

- *A priori* knowledge of scene relationships (e.g. the object must lie on the conveyor belt) is also expressible in algebraic form, allowing direct integration with observed and model relations.

We think that ACRONYM offers the best representation and machinery for geometric reasoning amongst the systems available.

Geometric constraints from surfaces

Most geometric reasoning operations (e.g. MERGE, TRANSFORM, INVERSE) can be defined abstractly, based only on the representation for positions. Hence, the key visual-geometry problems are relating object positions to image-model feature pairings (i.e. LOCATE and PREDICT).

Since we have been investigating using surface patches as the primary model and data descriptive primitive, the natural questions are:

- What geometric constraints can be obtained using these primitives?
- How can we represent them algebraically?

This section attempts to answer these questions.

Surface data

The raw data used here is three-dimensional surface information including absolute surface depth and orientation, such as found in Marr's 2½D sketch [6] or as used by Fisher [3]. The data is organized in a pointillist array aligned with a standard intensity image and segmenting the surface image into regions of nearly uniform shape, characterized by the two principal curvatures and the surface boundary. While there are no well-developed data acquisition systems that deliver high-quality surface descriptions of the scene, several promising data sources are under development, including laser striper range-finders, optical flow and stereo.

Some advantages of using surface information are:

- Surface information is explicitly three-dimensional, so three-dimensional properties can be calculated directly, rather than deduced from two-dimensional image properties.

- Surfaces can be segmented from both the model and the data according to the same criteria, hence producing directly corresponding features (disregarding scale questions).

- Features found on non-blocks-world objects (e.g. curved surface patches) are made explicit.

- The places where occlusion occurs (depth discontinuity boundaries) are identified.

Surface-based modelling

Model-based object recognition requires geometric object models. Here, the models [3,4] are designed for object recognition, not image creation, so the model primitives are based on matchable data features.

The models use surface patches as the major primitive, because the surface is the primary data unit (though space curves and blob-like volumes are also represented). This allows more direct pairing of data with models, comparison of surface shapes and estimation of model-to-scene transformation parameters.

It is assumed that the surfaces of a large class of objects can be approximately segmented into patches of nearly constant shape. The surface patches are then described by their principal curvatures and boundary. Surfaces have zero, one or two directions of curvature (positive or negative). Size and shape descriptions can involve variable quantities. General algebraic constraints involving variables can be included.

Objects are recursively constructed from surfaces or sub-objects using transformations of the coordinate reference frame. Each structure has its own local reference-frame transformation and larger structures are constructed by placing the sub-components in the reference frame of the aggregate. Variables occurring in the expressions that define the attachment relationship allow partially constrained relative placement.

One additional class of feature is the viewpoint-dependent feature, such as the tangential boundary on a cylinder. The cylinder has no such boundaries in its definition, but we generally observe these because of the viewer-object relationship. While the existence of these can be deduced from the geometric model, we assume here that the model represents them explicitly.

Other features of the models not discussed here are: viewpoint-dependent feature groupings and scale-based object representation simplifications.

Geometric constraints from model to data feature pairings

In ACRONYM [2] a single constraint is generated by first measuring the upper and lower bounds of some scalar quantity, S, and then relating these values to a known algebraic expression, E, which involves a subset, Q, of the quantifiers or variables of interest. If S_u and S_l are the upper and lower bounds on the true value of S then the contribution from the constraint would be the inequalities:

$$E(Q) \leq S_u$$

$$E(Q) \geq S_l.$$

ACRONYM, however, deals only with constraints generated from pairing two-dimensional edge features with a restricted vocabulary of model primitives. Here, we want to deal with richer three-dimensional descriptions for both images and models, and have found it useful to classify the types of constraints generated by pairings of the various image and model features.

All measurements from three-dimensional images take the form of vectors: point locations or directions in space. Constraints are then generated by pairing up the measured vectors with corresponding vectors in some model frame. When it is possible to make this pairing on a one-to-one basis we classify the constraint as type A. When there is ambiguity and we can only make the correspondence with some range of model vectors then the constraint is of type B.

Consider the example of a straight edge segment whose endpoints are obscured. The observed direction of the edge corresponds directly with its direction in the model frame and therefore generates a type A constraint on orientation. However, the visible length, l, is less than the true length l_0, and the visible endpoints do not correspond with the actual endpoints but with ranges of length $(l_0 - l)$ Consequently, the translational constraint generated from the endpoints is type B.

Constraints can be further divided according to the number, m, of point position vectors and the number, n, of direction vectors involved. For type A constraints, if $m = 0$ or $(m + n) < 3$ the model position is not fully constrained. Type B constraints, because they always involve some ambiguity, are never fully constraining. However, in any model-to-image pairing there will normally be several feature pairings and their degrees of freedom may well be orthogonal, so that their combined effect is fully to constrain the model position.

For dealing with errors we have adopted a scheme whereby point vectors have associated error spheres and direction vectors have error cones. An estimated point position is therefore given by four numbers: three for the location and one for the radius of the sphere containing the true point. Similarly, an estimated direction is associated with three numbers: two for the nominal direction and one for the cone angle containing the true direction.

Suppose that a type A constraint involves the pairing of model point $\mathbf{p}_m$ with image point $\mathbf{p}_i$, and model direction $\mathbf{d}_m$ with image direction $\mathbf{d}_i$. If $\mathbf{p}_i$ and $\mathbf{d}_i$ are error-free, then we can write:

$$\mathbf{R}(\mathbf{p}_m) + \mathbf{t} = \mathbf{p}_i$$

and

$$\mathbf{R}(\mathbf{d}_m) = \mathbf{d}_i,$$

where $\mathbf{R}$ is the rotational and $\mathbf{t}$ the translational part of the model position. If $\mathbf{P}_i$ has error sphere radius ε and $\mathbf{d}_i$ error cone angle μ, then the constraints would take the form:

$$|\mathbf{R}(\mathbf{p}_m) + \mathbf{t} - \mathbf{p}_i| \leq \varepsilon$$

and

$$\mathbf{R}(\mathbf{d}_m)\mathbf{d}_i \geq \cos \mu.$$

These two equations exemplify the constraint equations generated by the measurement of location and direction vectors respectively. Another similar equation is generated for each additional direction or point vector involved in a constraint.

The equational form of type B constraints is similar that that shown above except that $\mathbf{d}_m$ and $\mathbf{p}_m$ are parameterized by one or more variables. The variables are themselves subject to inequalities describing their allowed ranges. In the example of a straight edge with its ends obscured, the locus of model points corresponding to the visible end at $\mathbf{p}_i$ is a line which can be parameterized by a single variable β. We have (ignoring errors):

$$\mathbf{R}(\mathbf{p}_m + \beta\mathbf{d}_m) + \mathbf{t} = \mathbf{p}_i$$

and

$$0 \leq \beta \leq (l_0 - l).$$

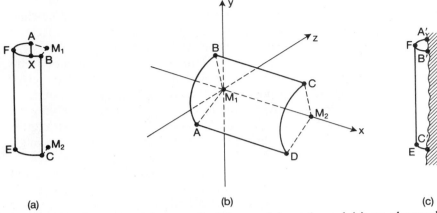

Fig. 8.1 (a) A data surface patch, (b) a model patch, and (c) an obscured data patch.

We now illustrate constraint generation with a practical example. Fig. 8.1(a) depicts a scene containing a cylindrical-shaped surface patch and Fig. 8.1(b) depicts the features recorded in the model. The model patch lies on a notionally infinite cylinder of radius r lying along the x-axis. Four segments bound the patch, two straight and two circular, meeting at the points labelled A, B, C and D.

We assume that the vision system has been able to identify the individual features of the patch so that the scene points A, B and C can be labelled as in Fig. 8.1(a) (the point D is not visible). There are then several constraints available from the data which together are more than enough fully to constrain the position of the patch.   Each visible boundary vertex, A, B and C, together with the tangents to the boundaries defining them, give a type A constraint involving one point location vector ($m = 1$) and two direction vectors ($n = 2$). The circular arcs AFB and EC imply the position of axis points m_1 and m_2 as well as the direction of the cylinder axis, yielding two type A constraints with $m = 1$ and $n = 1$. The lines BC and AX and the surface normals along them give similar constraints as does the obscuring boundary FE since we know that the surface normal along FE is perpendicular to the line of sight.

When many of the type A features are obscured we have to resort to type B features. In Fig. 8.1(c) we see that most of the patch is hidden behind an obscuring object (shaded).   None of the vertices are visible—instead we can only see where the boundary segments have been truncated at points A′, B′ and C′. As before, the obscuring boundary FE and the arcs A′FB′ and EC′ can be used to estimate the point positions M_1 and M_2 and the cylinder axis direction. This still leaves a rotational degree of freedom—the angle ϕ around the cylinder axis. To constrain this angle we can use the arcs A'FB' and EC' in a similar fashion to the example above of a line with its endpoints obscured. We then obtain type B constraints which narrow down the range of possible values of ϕ.

Every feature in a modelling scheme is capable of producing constraints which can be exploited by the geometric reasoner. We are currently engaged in compiling a catalogue of the constraint types available from the various features of the surface-based scheme described above. This will be used by the geometric reasoner automatically to relate any particular feature pairing to inequalities involving estimated quantities and free variables appropriate to that pairing. These inequalities will be the input to the constraint-solving engine described in the next section.

8.3   A network-based geometric reasoning engine

Since the geometric constraints defining object position are defined algebraically, following ACRONYM [2], one might use ACRONYM's constraint manipulation system (CMS). This CMS could bound expressions over non-linear inequalities and achieved respectable performance through a combination of case analysis and considerable symbolic algebra.

The advantage of using symbolic algebra instead of merely bounding each term is seen in the following example. Suppose we wish to bound the expression F_1F_2, where F_1 and F_2 are functions of x, which must lie in the range [1,2]. Suppose $F_1 = x$ and $F_2 = 2/x$. Bounding F_1F_2 by bounding and multiplying the individual terms gives the range [1,4]. However, using symbolic algebra to reduce the produce first gives the tighter bound [2,2].

Unfortunately, symbolic algebra is expensive computationally and we aim eventually to analyse scenes in real time. Hence, we are adapting it for use in a parallel evaluation network. ACRONYM's CMS is used to derive symbolic bounds on non-observable values (e.g. object orientation) in terms of observable values (e.g. direction of a surface normal), which would initially be expressed as variables. These bounds define a network of nodes representing expressions linked by their algebraic relationships. The network can be pre-compiled before any observations, with sections activated only when appropriate evidence is obtained.

The network is evaluated in parallel when image evidence is obtained:

• To provide bounds on position and size parameters.

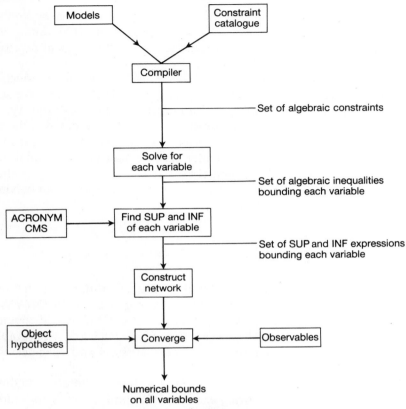

Fig. 8.2 Network definition from geometric constraints.

- To determine if an inconsistency is present (e.g. wrong model or model-data pairing).

The parallel evaluation provides the necessary performance.

   The methodology we are investigating is summarized in Fig. 8.2. At the top the constraint schemes discussed above are combined with models to define sets of algebraic constraints. Image observables are represented by variables at this stage. These constraints are then algebraically manipulated to produce a set of inequalities bounding each variable. An extended constraint manipulation system (CMS) based on ACRONYM's CMS is then used to form tighter bounds on each variable. The inequalities define the nodes and arcs in a network fragment, that are associated with object hypotheses to form complete networks (with inter-linkages through shared variables). When observable variables are bound to measured values the other variables (e.g. model or position variables) are forced into consistency.

At present, the definition of the network method is largely only a paper exercise. However, the example that follows later in the paper shows the feasibility of the approach, which we are continuing to investigate.

In the discussion below, the term **symbolic CMS** means our adaptation and improvement of ACRONYM. The term **network CMS** means the compiled network, which implements the same function as the symbolic CMS in a different manner.

Constraint Networks

A set of algebraic constraints can be viewed as a graph, where the nodes represent expressions and the arcs represent (binary) inequality relationships between them. Each node is identified with a processor and the arcs provide bounding input values.

We illustrate the network structure with the fragment for the inequality $a \leq b - c$ in the given variables. (Whether the variables represent model, position or observable values is not important.) Two other equivalent relations are $c \leq c - a$ and $b \geq a + c$. (In other words, we want to bound all variables by expressions in other variables.) The symbolic CMS is then used to calculate supremum and infimum expressions for some of these variables. This gives:

$$\sup(a) \leq \sup(b - c) \leq \sup(b) - \inf(c)$$

$$\sup(c) \leq \sup(b - a) \leq \sup(b) - \inf(a)$$

$$\inf(b) \geq \inf(a + c) \geq \inf(a) + \inf(c).$$

This formulation suggests introducing new nodes to represent expressions, and new processing elements for combining bounds.

A value node is now associated with each value, whether a variable or an expression, and represents the current supremum and infimum of the value. Hence the nodes relate to: $[a, b, c, b - c, b - a, a + c]$.

The computation for the supremum (i.e. sup (A)) picks the minimum of each bounding expression on the supremum, including the current supremum, because the bounds can never get worse (assuming a bound is always valid). Hence, if

$$\sup(A) \leq X$$

$$\sup(A) \leq Y,$$

then

$$\sup(a_{t+1}) \leftarrow \min(\sup(a_t), x_t, y_t)$$

is the updating function for the supremum of a.

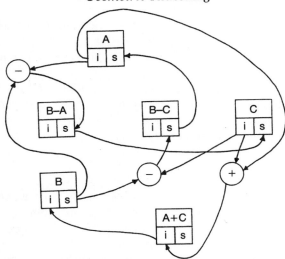

Fig. 8.3 Network fragment for $a \le b - c$.

Since the bounding expression for sup $(b - a)$ is a function of two expression bounds, we need to introduce arithmetic nodes into the network to compute these functions. (Because some constraints are dependent on the sign of the expressions, gating nodes are also used in the networks, but are not exemplified here.)

Applying these ideas results in the network shown in Fig. 8.3. This fragment is just for the single inequality given, and other constraints will introduce new nodes and linkages.

In execution, some variables acquire explicit bounds from measurements. The network then iteratively computes bounds on each expression until the network converges. The bounds give the allowable range for the variables consistent with all measurements and constraints. This implements the LOCATE and MERGE functions. Inconsistency is detected when sup $(F) <$ inf (F) for some expression F. The PREDICT function is implemented by examining the bounds on variables related to measurements, given bounds on other variables.

Extensions to ACRONYM's CMS

ACRONYM does not handle certain quadratic functions well. For example, with the following set of constraints, it fails to achieve the minimal upper bound. If

$$0 \le x \le 2, \ 0 \le y \le 2, \ \text{and} \ x + y \le 2,$$

then ACRONYM derives 4 as an upper bound on xy, whereas 1 is the correct upper bound. This weakness stems from not treating quadratic expressions as special cases.

The symbolic CMS we use is based on ACRONYM, but has been extended to give better bounds on functions of the form

$$ax^2 + bx + c$$

and

$$a \cos x + b \sin x,$$

where we have numerical bounds on a, b, c and x. For brevity, the extensions are not included here.

These extensions are particularly important because of the presence of rotation expressions in visual geometry, which often contain terms of the above form. (The $\cos x$ and $\sin x$ terms can be represented as y and $\sqrt{1 - y^2}$.)

Network pre-compilation

We have already seen how a series of algebraic constraints can be defined for different evidence types and model-data pairings. Though numerous, it is possible to enumerate all constraints on each structure before making any observations, except in the case where the number of data items is indeterminate (such as the number of surface normals observed for a given surface). Disallowing this case, the complete set of constraints on a model are represented with symbolic variables for each potential observed value. This constraint set would also include constraints defined as part of the model.

Whenever any measurements related to the object were made, then the related constraints could be enabled, or included in an active set. Alternatively, one could initialize all bounds on observed variables to be infinite.

Both defining the full constraint set and applying the symbolic CMS to produce symbolic bounding expressions can be done at model-compilation time. This may be slow, because there may be many variables in the expressions at this stage. However, this is unimportant, because this stage is done off-line. The only occasions for repeating the analysis are:

- New constraints types become understood and generated.
- New CMS techniques are implemented.

There is no difference in competence between the symbolic CMS and the network, provided data variables are set only to numbers. If the acquired values containing variables, then the symbolic CMS might be able to improve the bounds on the other variables. Because the data variables are intended to represent measured quantities, the competence assumption is reasonable. Hence, the only real difference between the two is performance, where using the network has clear run-time advantages because it can be executed in parallel using simple arithmetic units.

Using the symbolic CMS with some measured values might produce a simpler constraint expression than that containing measurement variables. For example, it may reduce

$$\sup (x) = \min(5, 10x)$$

to

$$\sup (x) = 5$$

where the constraint $x \geq 1$ is added. However, since the network executes in parallel, no significant execution time differences should occur.

Example

The geometric reasoning promoted in this paper is illustrated by an example of constraining position variables. The problem is simplified to two dimensions for illustration.

Fig. 8.4(a) shows a hypothetical unit line segment. Its location is defined by the translation of its central point and orientation (anticlockwise) about the central point. The model segment is (redundantly) characterized by its position in a local coordinate frame $\mathbf{p}_m = (0, 0)$, its direction $\mathbf{d}_m = (1, 0)$ and its normal $\mathbf{n}_m = (0, -1)$.

Suppose the model appears in the scene with the rotation θ and translation $\mathbf{t}_d = (t_x, t_y)$ (see figure 8.4). Here, $\theta = -\pi/4$ and $\mathbf{t}_d = (0, 0)$. Hence, the scene position, direction and normal are:

$$\mathbf{p}_d = \mathbf{R}_\theta \mathbf{p}_m + \mathbf{t}_d = (t_x, t_y) = (0, 0) \tag{8.1}$$

$$\mathbf{d}_d = (0.7, 0.7)$$

$$\mathbf{n}_d = (-0.7, -0.7),$$

where $\mathbf{R}_\theta$ is the rotational matrix

$$\begin{bmatrix} \cos \theta & -\sin \theta \\ \sin \theta & \cos \theta \end{bmatrix}.$$

Suppose next that we observe an estimated normal $\hat{\mathbf{n}}_d = (\hat{n}_x, \hat{n}_y)$ at some unknown point $\mathbf{v}_d$ on the segment

$$\mathbf{v}_d = \mathbf{p}_d + l\mathbf{d}_d, \tag{8.2}$$

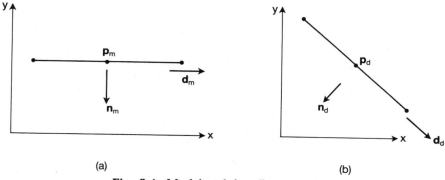

Fig. 8.4 Model and data line segment.

and hypothesize that $l \in (-0.4, 0.4)$. This is a type B constraint (see Section 8.3) because we do not know which model point exactly corresponds with the point we measure.

Let the observed location be $\hat{v} = (\hat{v}_x, \hat{v}_y)$. Since our observations must have some error, we know that

$$|v_d - \hat{v}| \le \varepsilon_1 \qquad (8.3)$$

and

$$1 - \hat{n}n_d \le \varepsilon_s. \qquad (8.4)$$

Assume $\varepsilon_1 = 0.05$ and $\varepsilon_2 = 0.005$.

Constraint equations 8.1 − 8.4 above define the position of the segment. The reasoning task is to combine the constraints to bound the possible range of its positions. Constraints 8.1 and 8.2 are part of the model, whereas 8.3 and 8.4 are returned by the LOCATE function. Representing all constraints in the network implements the MERGE function.

Hence, our problem is to bound the position variables θ, t_x, t_y and l, given the observed variables $\hat{n}_x$, $\hat{n}_y$, $\hat{v}_x$ and $\hat{v}_y$. Assume that we observe the true normal at the true central point. Then:

$$\hat{n} = (-0.7, -0.7)$$

$$\hat{v} = (0, 0).$$

The geometric constraints 8.1 − 8.4 reduce to a set of algebraic constraints on the variables, such as:

$$t_x \le 0.05 - l \cos \theta$$

$$t_x \ge -0.05 - l \cos \theta$$

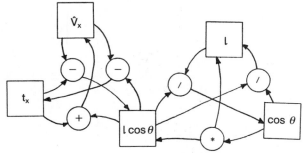

Fig. 8.5 Network fragment for 'X'-linked nodes.

$$\cos \theta \leq -1.34 - \sin \theta$$

$$l \leq 0.4$$

plus a general equality:

$$|\cos \theta| = \sqrt{1 - \sin^2 \theta}.$$

These together define the network solving for the position variables. Fig. 8.5 shows the fragment for the x and $\cos \theta$ nodes in schematic form and embodies the relations between $\hat{v}_x$, t_x, 1 and $\cos \theta$. From constraint equations 8.1, 8.2 and 8.3 we know that

$$|\hat{v}_x - l \cos \theta - t_x| \leq \varepsilon_1.$$

The initial ranges of the observed variables are $\inf (\hat{v}_x) = -\varepsilon_1$, $\sup (\hat{v}_x) = \varepsilon_1$, $\inf (l) = -0.4$, and $\sup (l) = 0.4$.

Letting this network converge, the following bounds on the position variables are obtained:

$$-0.4 \leq l \leq 0.4$$

$$-0.88 \leq \theta \leq -0.68$$

$$-0.36 \leq t_x \leq 0.36$$

$$-0.36 \leq t_y \leq 0.36.$$

Since this is equal to the best bounds obtained by analytic solution, the network has been successful at bounding the variable values.

Network behaviour

Three key computational questions are

- How does the network get evaluated?
- Does it converge?
- Does it converge to the best bound?

The network is intended to be executed in parallel, with each node in a separate processor.   It could be evaluated synchronously or asychronously in a MIMD (multiple instruction, multiple data stream) processor with non-local connectivity, because expression relationships are not regular. Since the expression relating two variables is typically contains fewer than 20 terms, and since it appears to only take a small number of iterations (e.g. $5-10$) to evaluate the networks (at least the simple ones we have tried so far), we expect fast convergence.

It is easy to show that the networks must converge, that is, not oscillate. At any time the current bounds on a value V must be true. Then, if an upper bounding expression increases in value, the original bounds on V must still hold, as it then makes no sense to increase the range of potential values for V. Hence, the bounds can only get tighter, though perhaps asymptotically.   As the bounds can only be equal (inconsistency is declared if they cross), each bound has a limit, so must converge.

Here, parallel evaluation is simulated serially using conventional programming (i.e. Prolog). Keeping track of which dependent nodes need re-evaluating when one node changes eliminates recomputation in stable sections of the network.

ACRONYM's CMS was optimal when producing numerically bounded variables (i.e. not resulting in expressions in other variables) over sets of linear constraints, based on work by Bledsoe [1] and Shostak [9]. Since we reproduce all alternatives of the symbolic reasoning in the network, without substituting values for the data variables until later, it is likely to have the same performance on linear constraint sets, but we have not proved this result.

As most of the constraints derived in the previous subsection are non-linear, we cannot expect optimality. However, the inclusion of the quadratic and trigonometric bounding extensions should improve performance by accounting for most of the non-linear cases.

Continuing research

Some areas where we are continuing research are:

Trigonometric wrap-around

The example was set up so that we did not have to worry about cases where $\cos \theta > 0.9$ or $\cos \theta < -0.9$.

Bounding x in ax < b when the sign of a is unknown

If a can take values in the range $[-r, s]$, then b/a can take any value. Hence, this constraint is not useful until the sign of a is known. Thereafter, the sense of the constraint depends on the sign. Consequently, the network structure should probably allow gating to switch on constraints when parity conditions hold.

Automatic network compilation

The network should be automatically constructed from the constraints. Because the geometric constraints can be defined in standard form, isolating the variables, applying the symbolic CMS and compiling the network should be easy.

8.4 Conclusions

This paper has discussed three topics relating geometric reasoning to computer vision:

- What tasks must a visual geometric reasoner perform and what functions are needed to implement them?

- What classes of constraints exist from pairing three-dimensional model features to three-dimensional image features, and how they can be represented?

- How can constraints be represented in a computational network to provide fast parallel bounding of variables and expressions, which is necessary for real-time estimation of object position and model parameters?

Through analysing several high-performance vision systems, the main reasoning tasks identified were: forming position constraints, combining constraints, detecting inconsistencies, transforming positions, variable binding and feature prediction. The main functions needed to implement the tasks were specified as abstract data types.

Sets of equality and inequality constraints were based on pairing point and vector features defined three-dimensional model and data features. The constraints were represented in algebraic form. A network structure could be defined based on inequalities derived from the constraints. This network is suitable for parallel evaluation. An example of a network that successfully performing geometric reasoning was given.

This research is being pursued as part of a larger project on model-based visual scene understanding, which is still being implemented. Consequently, details of network reasoner performance on larger problems are not yet available.

Acknowledgements

The work presented above was funded under Alvey grant GR/D/1740.3. Thanks go to J. Aylett and the Robot/Vision discussion group for helpful comments.

References

1. W.E. Bledsoe, 'The sup-inf method in Presburger arithmetic', Memo ATP 18, Department of Mathematics and Computer Science, University of Texas at Austin, 1974.

2. R.A. Brooks, 'Symbolic reasoning among 3-D models and 2-D images', *Artificial Intelligence* **17** (285-348), 1981.

3. R.B. Fisher, 'From surfaces to objects: recognising objects using surface information and object models', PhD Thesis, University of Edinburgh, 1986.

4. R.B. Fisher, 'SM: A suggestive modelling system for object recognition', University of Edinburgh Department of Artificial Intelligence Research Paper 298, presented at the 1986 Alvey Computer Vision and Image Interpretation Meeting, 1986.

5. J.V. Guttag, E. Horowitz, and D.R. Musser, 'The design of data type specifications', in *Current Trends in Programming Methodology IV* (R. Yeh, ed), Prentice-Hall, Englewood Cliffs, NJ, 1987.

6. D. Marr, *Vision*, Freeman, 1982.

7. M.J.L. Orr and R.B. Fisher, 'Geometric reasoning for computer vision', University of Edinburgh Department of Artificial Intelligence Research Paper 311, presented at the 1986 Alvey Computer Vision and Image Interpretation Meeting, 1986.

8. R.J. Popplestone, A.P. Ambler and I.M. Bellos, 'An interpreter for a language describing assemblies', *Artificial Intelligence* **14**, 1980.

9. R.E. Shostak, 'On the sup-inf method for proving Presburger formulas', *Journal of the ACM* **24** (529-543), 1977.

Discussion

(This paper was presented by Fisher)

Knapman: I'm a bit concerned that you are depending on a quality of data from the vision system that it may not be able to deliver. For instance, determining surface normals from an actual scene—that is to say, normals across a surface—is a process of very variable reliability. Its accuracy depends on nice texture and so on; you may have a smooth surface with just a bit of information from a specularity in the middle and the lines around the edges. Do you have any comments on that?

Fisher: There are two possibilities here. The first is to have a very large tolerance, and to assume that, for almost all of the time our information is going to be within that ε, and use that value in the reasoning system. The second is to determine tolerance values as part of actually producing the constraint in the visual reasoning system, not the geometric reasoning system. For instance, if a constraint emerges from a textured region, orientation estimates are likely to be good, and so that constraint will be used, or used with a very tight ε. If a constraint is not quite so clearly valid, it would be used with a larger ε, or not at all. That's really more of a visual question than a geometrical reasoning question; however, your point is well taken.

Wright: Can you tell me whether the process you describe only works on isolated objects? If you've got multiple objects in a scene which might, for instance, actually overlap fragments from more than one object, how do you then differentiate between these constraints and avoid applying an inconsistent set of constraints to your network?

Fisher: Again, I would say that is largely a visual problem, not a reasoning problem, in the sense that your systems should be responsible for correctly pairing the constraints. However, you can use the geometrical reasoning network for testing whether the constraints have been paired consistently. Ordinarily, I would expect a vision system to identify all the features in a scene, perform a pre-grouping of the features and relate them to a model. Geometrical reasoning is one of the constraints for testing whether that pairing is correct. There are other constraints that one can use on the pairing of features beforehand, so it's not necessary to depend upon this strictly.

Wright: You can't identify inconsistent constraints in your set? You can't separate the 99% of constraints that match up from the 1% that don't?

Fisher: The way the network is set up now, I would say, probably not. The bounds on an expression are a variable cross. You should always have a sup larger than the inf, but if they ever cross, than you have an inconsistency somewhere in your network, then this inconsistency could have arisen from anywhere in the network, so it's hard to identify its source. I would say at the moment that we probably cannot solve that problem.

One approach is incrementally to add your constraints or to activate constraint fragments in the network. If the first constraint converges, it's consistent. Add the second constraint and that converges; the bounds are tighter, but things are still consistent. If the third piece of evidence is suddenly inconsistent, that doesn't say that the third one is wrong, it's possibly one of the first two, but it narrows things down a bit. To implement that would be a matter of the control structure, I think. It's a good question—we really haven't thought of that problem yet.

9 A representation for geometrically toleranced parts

A. D. FLEMING

9.1 Introduction

During engineering design of a part, a tolerance specification must be attached to the part in order to state what variations in its shape are acceptable.   However, it is not easy to decide whether a proposed tolerance allocation is satisfactory.   Therefore, it would be useful to automate the analysis of tolerances. This paper described how a part with geometric tolerances can be represented computationally.

To understand some of the difficulties it is useful to consider the process of machining a part. A feature of a part may be cut or drilled while the part is being supported by some other feature.   Different supports may be used for different cutting operations. A feature ends up being positioned with a known accuracy relative to the current supporting feature but the accuracy of its position relative to some other feature is not known directly.   These unknown relationships may be important for the satisfactory working of the part and so it is useful to be able to verify that they are within required bounds.

The standards used by engineers to specify tolerances are contained in B.S. 308 [1]. One way of specifying tolerances is to indicate upper and lower bounds for given dimensions of the part.   However, dimensions are not always definable on a real part. This is because the surface of a manufactured part is not perfectly formed. It is impossible to define a unique distance between two imperfect surfaces. Techniques called **geometric tolerancing** enable imperfect forms to be taken into account. Use is made of **tolerance zones** which are regions in which a feature of the real part must lie. A tolerance is specified by giving the properties of the zone in which the part of the surface of the part must lie.

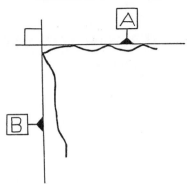

Fig. 9.1

A formalization of tolerances

The work presented in this paper is based upon formalization made by Requicha [3] of geometric tolerancing techniques. A brief summary of his formalism is made here as it is essential for understanding of the rest of the paper.

A **nominal feature** is defined as a region taken from the surface of a real part. Features may be classified as being **simple** or **composite**. A simple feature has a simple geometric shape such as a plane or a cylinder and a composite feature is a set of simple features.

A **datum** is defined as an infinite plane, an infinite straight line or a point embedded in a part. Any nominally symmetric feature can define a datum. A datum is located on a real feature using the so-called **measuring procedure**. A copy of the nominal feature is taken and is scaled and oriented so that it just contains the real feature. This copy of the nominal feature is called the **measuring solid**. Its orientation is chosen so that its size is minimized. The appropriate plane, line, or point of symmetry of the copy of the nominal feature is taken as the datum for the real feature. In the case of planar features the datum should just rest against the surface of the feature.

A **system of datums** is a set of datums which may be ordered. In Fig. 9.1 datum A is primary and datum B is secondary. A is parallel with the overall trend of its associated feature whereas B is constrained to be correctly positioned (in this case perpendicular) relative to A and as a result only manages to touch its associated feature at one end.

There are various types of tolerance but for simplicity this paper deals only with tolerances of position, size and form. All these are the defined by tolerance zones. In the case of a cylindrical feature the tolerance zone is an infinitely long cylindrical shell as shown in Fig. 9.2. It has three characteristics—position, size and thickness—which are

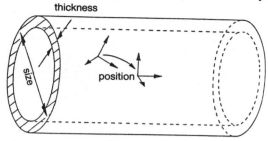

Fig. 9.2

specified or unspecified depending on the type of tolerance. In the case of a cylindrical zone its 'size' is its diameter. The properties of zones of different tolerance types are as follows:

- A **position tolerance zone** has fixed size and thickness and fixed position relative to a specified system of datums.

- A **size tolerance zone** has fixed size and thickness but undefined position.

- A **form tolerance zone** has fixed thickness but undefined size and undefined position.

- For the real feature to satisfy a tolerance it must be possible to enclose the real feature with a tolerance zone with the correct properties.

A complication with tolerance specifications is that networks of zones and datums arise. Features make use of datums in the definition of their position tolerances and these datums are defined by other features and so on. A network can be formed with arcs representing relationships and nodes representing features and datums. This paper shows how the network can be represented and how constraints can be associated with its arcs. It discusses the geometry of the different situations in which constraints arise.

Example

An example is presented here to show how a network of features and datums is obtained from a tolerance specification of a part. The part shown in Fig. 9.3 is a plate with two groups of four holes. The function of the holes is to attach two dials and so the holes in each group must be positioned accurately if they are to meet up with holes in the dials. However the relative position of the two dials is not critical and so the

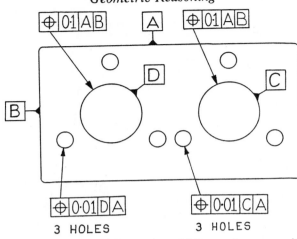

Fig. 9.3 The symbol ⊕ indicates a position tolerance with the given parameter relative to the given datums.

relative position of the two hole groups need not be defined so accurately.

An ordered datum system, A and B, is defined by two sides of the plate (Fig. 9.4). The two last holes are positioned with respect to datums A and B with a relatively large tolerance of 0.1. Each defines a line datum, denoted by C and D, corresponding to its axis of symmetry. There are three small holes round each large hole with a position tolerance of 0.01. Three of the small holes are positioned relative to the datum-system of primary datum C and secondary datum A. The distance of their position tolerance zones from C is fixed and the angle of the line from C to a zone must be correct relative to the orientation of A. The other three small holes are positioned similarly but relative to datums D and A.

Consider the exaggerated illustration of the real part in Fig. 9.4. As a simplification only one set of holes is shown. Datums and tolerance zones are included and the actual surfaces of part are shown by bold lines. Notice that the tolerance zones are at the correct position relative to datums which are located by the appropriate real features. Real features are contained in the zones.

A network can be created, as shown in Fig. 9.5, with circle nodes representing features and cross-nodes representing datums and arcs to show which pairs of items (features and datums) have constrained positions. Directed arcs show that one item defines another. Every datum is defined by a feature and every feature (except for the sides of the plate represented by the top two nodes in the network) had its position defined relative to one or more datums. The six nodes at the bottom represent the small holes. Note that datums and features have

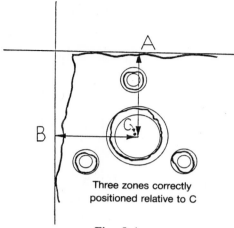

Three zones correctly
positioned relative to C

Fig. 9.4

been kept distinct in the network. This is consistent with the formalism but is not obvious from an engineering drawing. The meaning of nodes representing features and the meaning of arcs is not yet clear, however.

In this example there are three types of relationship between datums and tolerance zones:

- Datums in a datum system are correctly positioned with respect to one another.

- Tolerance zones are correctly positioned with respect to specified datums.

- Datums are located by their associated features and so are approximately aligned with the corresponding tolerance zones.

9.2　Formalizing the problem

These relationships put constraints on the positions of tolerance zones and datums. Before discussing relationships further there are several topics that need to be formalized.

Design requirements

An important assumption being made in this paper is that a tolerance specification is only useful if all copies for the part that it defines are functional. This is because partial results are only of use in the presence of statistical information and the statistics of tolerances are here

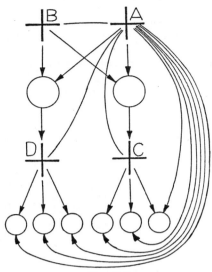

Fig. 9.5

ignored. Although, for example, it might be satisfactory if only 99% of parts manufactured function properly and for there to be 1% wastage it is impossible to know, without statistical information, what percentages occur. Hence, all useful measurements and calculations must apply to 100% of manufactured parts. Any design requirement can be expressed by saying that the result of a measurement must be within a certain range.

Computational representations

Using the concepts of Requicha's formalism described above, a mathematically precise definition of the semantics to tolerances can be constructed. Something must be said about the validity of a computational representation and how it relates to the formalism.

The set of allowable shapes of a part is called its **variational class**. It is a set of subsets of R^3. A member of a variational class will be called an **instance** of a part. The formalism presented above, along with a tolerance specification for a part, provides a mathematical description of a variational class. This paper shows how the variational class can be described in a computer. This description will be referred to as a **computational representation**. Ideally, the variational class described by a tolerance specification and the variational class described by the computational representation are the same. However, some differences can be tolerated.

The validity of a computational representation can be checked by determining if it can be used to answer questions correctly about the entity that is represented. Questions asked in the real world are **measurements** and questions asked about the representation are **calculations** or simulated measurements. Note that, whilst a measurement performed on a single instance of a part returns a single value, a calculation on a representation of the set of possible parts returns the set of all values that could be obtained from the measurement.

It is not always possible to arrange that calculations return exactly the same set of values that could be obtained from measurements. This is because either the representation may be an approximation or the calculation may be an approximation.

The type of approximations that can be tolerated can be established as follows.

Let V be the variational class of a part. Let m be a measurement so that i is an instance of the part then $m(i)$ is the value obtained by the measurement if it is applied to i. Let $m(v)$ be the set of all possible values of the measurement, i.e. $m(i) : i \in V$.

Suppose that we require the result of a measurement, $m(i)$, to be within a certain range. Then it follows that all members of $m(V)$ must be in this range. Therefore, a design requirement involving dimension m can be written, for some set M, in the form:

$$m(v) \in M.$$

Suppose we have an inaccurate representation which defines variational class V' and measurement m' instead of V and m. We would like to be able to deduce $m(C) \in M$ from $m'(V') \in M$. This can be done if it is ensured that all approximations cause $m'(V')$ to contain $m(V)$. This gives a criterion for deciding whether approximations are acceptable.

Representation of zones and datums

The computational representation of a toleranced part can be achieved using three types of geometric entity. These are:

- Nominal features—subsets of the surface of the part. Simple features have simple geometric shapes such as planes and cylinders. Composite features are composed of sets of simple features.

- Tolerance zones—defined by attaching attributes to simple or composite features.

- Datums and systems of datums. A simple datum is a point, an infinite line or an infinite plane. A system of datums is a possibly ordered set of simple datums.

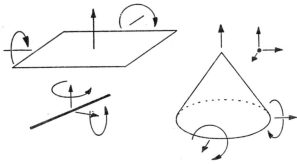

Fig. 9.6

The position of a zone or a datum is defined by size variables. They represent displacements of the items from their nominal positions (the positions of the nominal features to which they correspond). Translations are measured along the axes of some specified coordinate system and rotations are measured about the axes. These variables will be called degree-of-freedom variables, or **DOF-variables.** Constraints on the positions of items will be expressed as equalities or inequalities involving DOF-variables.

Note that most zones and datums have fewer than six degrees of freedom. This is because many motions map the items onto themselves and as a result can be ignored. Fig. 9.6 shows examples of zones and datums with their degree of freedom; an infinite plane has two rotations and one translation; an infinite cone has two rotations and three translations; a point has three translations and no rotations. Hence, a tolerance zone has some DOF-variables which are **redundant.**

Zone-datum structures

The positions of the zones and datums of a part are constrained relative to one another in ways that will be described later in this paper. The set of zones and datums in definite positions can be thought of as a rigid structure. It will be called a **zone-datum structure** or ZDS. There is a set of configurations of zones and datums for which the constraints on the positions of zones and datums are satisfied and this will be called the **ZDS set.**

Extent-solids

Informally, the **extent-solid** of a feature is a solid which is bounded in the same way as the extent of the nominal feature but extends infinitely in directions along normals to the nominal feature. An extent-solid of

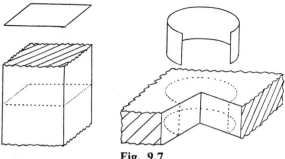

Fig. 9.7

a feature is a convenient way of describing the extent of the feature. It can be used to isolate the portion of a tolerance zone which corresponds to be extent of the nominal feature. Fig. 9.7 shows two features and their extent-solids.

The extent-solid of a feature must have the following properties:

- It contains the nominal feature.
- Its surface contains the boundary of the nominal feature.
- Where its surface intersects the boundary of the nominal feature its surface is normal to the nominal feature (Fig. 9.8).

In the case of a planar feature the extent-solid is an infinite prism with a cross-selection which is the same shape as the nominal feature. In the case of a cylindrical feature the extent-solid is an infinite slab with the thickness the same as the length of the cylinder. If the cylinder is not complete then a corresponding segment is removed from the extent-solid (see Fig. 9.7).

An extent-solid can be used to define the **significant portion** of a tolerance zone. It is defined as that part of the zone which lies inside the extent-solid. Although most tolerance zones have infinite extent, a real feature only 'occupies' a finite portion of a zone. The extent of a zone can be reduced without significantly affecting the class of real features which it allows. Actually, the significant portion only approximates the portion occupied by the real feature. However, this approximation is small enough not to cause any problems.

Note that a zone divides the volume inside an extent-solid into three regions (Fig. 9.9). It is useful to consider the distribution of material and air in these three regions. The region inside the zone contains a mixture of air and material.

One of the other regions contains entirely air near to the zone whereas the remaining region contains entirely material near to the zone. What happens far away from the zone is undetermined.

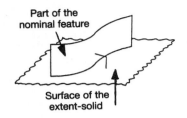

Fig. 9.8

Signed distances

Some constraints between tolerance zones and datums will be obtained using signed distances. Given two regions of space, A and B, the **signed distance** between them is denoted by sdist(A, B) and is defined as follows:

- When A and B do not intersect it equals the distance between the closest points of A and B.

- When the interiors of A and B intersect it is negative and equals the negative of the shortest distance that one region would have to be moved to separate the regions.

- When only the surfaces of A and B intersect it equals zero.

Despite the three-part definition the signed distance is a continuous function of the relative position of the regions. It allows geometric constraints to be expressed algebraically. For example, the geometric condition that two regions do not intersect can be converted to an algebraic constraint:

$$\text{sdist}(A, B) \geq 0.$$

If the positions of the regions are expressed in terms of DOF-variables then this is an expression involving DOF-variables.

9.3   Relationship between zones and datums

A network can be constructed with zones and datums as nodes and with relationships as arcs. A relationship constrains the positions of the items which it connects. The four ways in which geometric relationships between zones and datums occur are listed below. All except the third occur in the example in the Introduction.

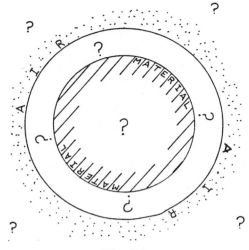

Fig. 9.9

- A position tolerance zone can be 'correctly positioned' relative to some datum.

- The datums in a datum system can be 'correctly positioned' relative to one another.

- Tolerance zones associated with the same feature have constrained positions.

- A datum is constrained by the measuring procedure relative to the zones of its associated feature.

Each gives rise to a geometric constraint which is converted to an algebraic constraint (a set of inequalities). Only simple features are dealt with in this section: composite features are left until the next section.

Relationships locating zones relative to datums

This type of relationship occurs when a zone of position or orientation tolerance has its location constrained by a datum. In the example in the Introduction the tolerance zones of the large holes were located by datums A and B. Therefore, the zones of position tolerances of these features make a relationship with both of A and B. Each relationship removes some degree of freedom from the relative position of the datum and the zone.

It is often convenient to derive the constraints relative to coordinate systems other than the main coordinate systems of the zones or datums.

Then the constraints can be converted by making variable substitutions derived from the transformation between the convenient coordinate system and the central coordinate system.

It is always possible to choose a convenient coordinate system so that the constraints from this type of relationship take the form of DOF-variables equated to zero. This is because the nominal value of a DOF-variable is taken as zero. The chosen coordinate system (attached to the zone) is aligned with the coordinate system of the datum.

Relationships between datums in a datum-system

A datum-system is a set of datums which may be ordered. Ordered datums are located relative to one another in precisely the same way that their associated nominal features are located. In the example in the Introduction the two planar datums are perpendicular because their associated features are perpendicular. The angle between the datums is fixed and so rotation of the secondary datum about an axis coincident with the line of intersection of the datums is constrained. However, all other degrees of freedom are unconstrained. As with the previous two relationship types, coordinate systems can be chosen so that the constraints equate DOF-variables with zero.

The degrees of freedom which are constrained depend on the types of datums involved and the disposition of the datums. For example, an ordered datum-system consisting of two perpendicular intersecting line datums would cause one rotational degree of freedom to be constrained. However, a datum-system consisting of two parallel line datums would cause one translational and two rotational degrees of freedom to be constrained.

The previous two relationship types involved equality constraints but the following two involve inequalities. The extent of a feature is important in these relationships.

Zones associated with a single feature

Formally, the location of zones of size and form tolerance are completely undefined. However, there is a simple geometric consideration that does constrain their locations: there must be room in the intersection of the zones for a real surface.

Consider a planar feature with a size tolerance and a position tolerance. Fig. 9.10(a) shows the zones at their extreme inclination given that a surface with the required extent has to fit inside them. Any larger inclination would cause the length of the region of intersection to

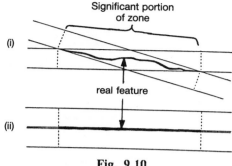

Fig. 9.10

be shortened. Fig. 9.10(b) shows the extreme translational displacement that can occur if the zones are to intersect.

Constraints on the location of a form-tolerance zone can be found in a similar way. However, in addition there is a constraint on the size of the zone.

The constraints on the relative position of two zones can be derived from the fact that the totally material region of one zone must not intersect the totally air region of the other zone. This condition is illustrated in Fig. 9.11(a) − (c) for a planar feature. In Fig. 9.11(a) the zones are shown with the material region of one zone intersecting the air region of the other. In Fig. 9.11(b) the air region of the first intersects the material region of the other. Both of these cases are impossible and the region of intersection is not as long as the significant portions of the zones. Fig. 9.11(c) shows a situation where the condition is satisfied.

Algebraic constraints on the locations of the zones can be derived from the fact that the signed distances between the totally material region of one zone and the totally air region of the other must be greater than zero. For a system to derive these constraints it would have to know how the relevant signed distance can be expressed in terms of DOF-variables; it would have to know this for every type of feature and every shape of feature extent that occurs. Either it would have access to an algorithm that could evaluate a signed distance for any feature shape or to a catalogue of expressions for different feature shapes. In some cases it can be shown that the constraints obtained by these methods are the best available though there are many situations where they are not. The constraints do however provide a useful bound on the set of possible positions.

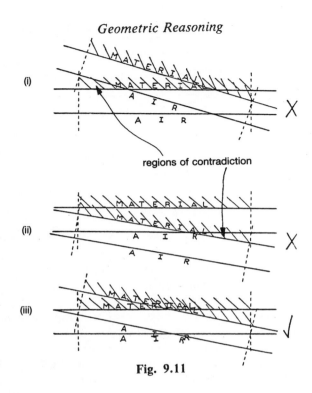

regions of contradiction

Fig. 9.11

Relationship between a datum and a zone

Every datum is defined by a planar feature or a symmetric feature which may be simple or composite, though composite features are not considered until the next section. In the example of Figs. 9.4, 9.5 and 9.6, datums C and D were defined by the large holes and so made relationships with the zones of these holes.

Fig. 9.12 shows that a line datum centred on a cylindrical feature is approximately parallel with the zone and its approximately coaxial with it in the significant portion of the zone. The deviation of the datum from the axis of symmetry of the zone depends on the thickness of the zone.

More precisely, the position of a datum relative to a symmetric feature of an actual part is constrained geometrically by the 'measuring procedure' defined earlier. It is not immediately obvious how this geometric constraint can be converted into an algebraic form.

Note, though, that there are two geometric conditions that the measuring solid must satisfy:

- The maximum expansion ever required of the measuring solid is such that it can enclose the size tolerance zone. In this situation the

Significant portion
of zone

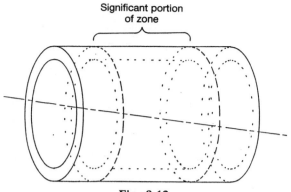

Fig. 9.12

surface of the measuring solid is coincident with the surface of the zone that can be guaranteed to be next to air.

• The measuring solid always encloses the totally material region of the zone. therefore, it encloses the portion of the surface of the zone that lies inside the extent solid and which is next to material.

Fig. 9.13 shows a section through a cylindrical zone with the measuring solid at the most extreme inclination that satisfied the above constraints. The diameter of the measuring solid is equal to the outer diameter of the zone and it contains the surface of the significant portion of the zone which is next to material. The second condition can be expressed in terms of signed distances. It says that the signed distance between the totally material region of the zone and the complement of the measuring solid is greater than zero. The size of the measuring solid should be assumed to be as large as possible in order to obtain the most general constraints. The constraints on the position of the datum are the same as the constraints on the position of the measuring solid.

It is not obvious that these methods result in constraints which are as strong as possible. However, the following example will show that for some types of feature, at least, the constraints obtained are realistic.

Fig. 9.14 shows the same zone as Fig. 9.13 containing a real feature. The real feature has been chosen so that slightly more than half of its surface is inclined and the other half is parallel with the zone. A measuring solid enclosing this surface attains minimum size if it is parallel with the inclined part of the surface. (Note that, if less than half of the surface has been inclined then the measuring solid would attain minimum size by being parallel with the zone.) The inclination of the datum implied by this surface is that same as the maximum inclination derived from the geometric constraints above. Therefore, for a cylindrical feature, the constraints derived from the geometric conditions are accurate.

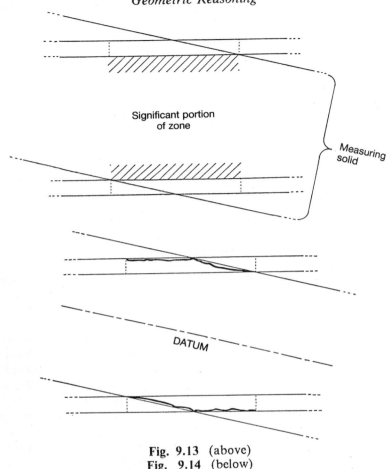

Fig. 9.13 (above)
Fig. 9.14 (below)

What happens when there are several tolerances associated with the same feature? Suppose that the datum is constrained relative to each zone and the zones are constrained relative to one another forming the network in Fig. 9.15. It would be convenient if the total constraints on the datum could be deduced from the constraints in this network. Unfortunately, the total constraints can be stronger than this.

Consider a planar feature subject to form and position tolerances. Fig. 9.16(a) shows a section through the two zones containing a real surface. The relative inclination of the zones and the shape of the real surface have been chosen to give the most extreme inclination of the datum. If the zone of form tolerance were at a greater inclination then the length of the intersection of the zones would be reduced below the length of the nominal feature. No surface that fits in the region of intersection can give rise to a datum which is more inclined. Therefore,

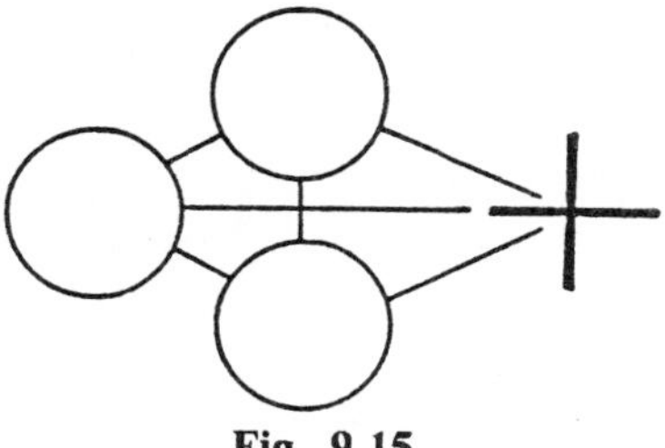

Fig. 9.15

this datum has the largest inclination that can be obtained. Fig. 9.16(b), on the other hand, shows the position tolerance zone on its own and again a surface has been chosen to give the maximum inclination of the datum. As in the example above, slightly more than half of the real surface is inclined and so this determines the datum's inclination. In Fig. 9.16(c) the same is done for the form tolerance zone. Figs. 9.16(b) − (c) are illustrations of the constraints that would be obtained by the methods described above applied to the individual zones. Fig. 9.16(d) illustrates the combination of these constraints. The datum would attain the lesser of the two inclinations shown. However, it is obvious from Fig. 9.16(a) that the actual maximum inclination is actually less than either of these. Hence, the constraints derived from the network in Fig. 9.15 are not maximally strong. It can be guaranteed, however, that they are weaker than reality because the constraints in each arc is correct. The existence of better techniques is not obvious.

So far the discussion of this section has dealt with datums defined using a measuring solid. In the case of planar features, however, the datum is simply rested against the real surface. For these it is necessary to use inspection to decide what are the worst possible inclinations and displacements that a datum could attain relative to a tolerance zone. Inspection could also be used to deal with a few simply shaped, commonly occurring symmetric features such as cylinders, cones and spheres. In this way more accurate constraints would be obtained that those obtained by the general technique. The constraints obtained in this way would be catalogued as case-by-case rules. For example, Fig. 9.17 shows worst-case inclinations and displacements for a planar feature and a planar datum. Simple analysis of the geometry allows constraints to be found on the position of the datum relative to the zone.

Secondary datums

The position of a secondary datum is also constrained relative to the tolerance zones of its associated feature but the constraints are weaker. This is obvious in Fig. 9.1 where the primary data is forced to be

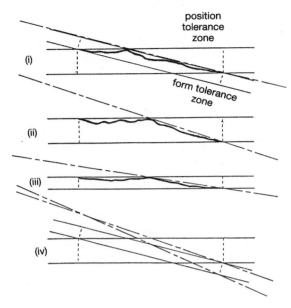

Fig. 9.16

parallel with the overall trend of its associated feature and to touch the feature, but the secondary feature only touches its associated feature at one point.

As discussed in Section 9.3, the relationship between a primary datum and a secondary datum removes some degrees of freedom. The variables associated with these degrees of freedom must not occur in the inequalities associated with the relationship that the secondary datum makes with its tolerance zone. All other DOF-variables must be constrained in the same way that they would be if the datum were primary.

Approximations in constraints

The inequality constraints obtained in the last two types of relationships described above contain approximations. Approximations may be acceptable for two reasons. Either they are small enough to cause no important effect or they result in an overestimate of the class of zone-datum structures (and hence an overestimate of the variational class). The latter is a slackening of the constraints. Approximations occur in three ways.

- The extent of a real feature in approximated to the extent of a nominal feature. Wherever a value representing the extent of a

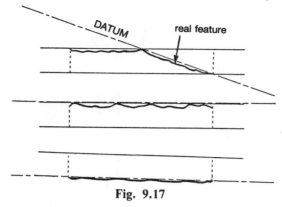

Fig. 9.17

feature appears in one of the inequalities it should actually be a variable bounded according to the variation in size of the feature. However, it has already been noted that this type of approximation can be ignored because tolerances are much smaller than the extent of features.

- It may be necessary to approximate the extent of a feature if its extent is a very complex shape. This is because the resulting inequalities would be too complex to be dealt with by the system. By taking an underestimate of the extent the constraints are slackened.

- The techniques for finding constraints produce constraints that are weaker than would occur in reality.

The example again

Fig. 9.18 shows the complete formalized network for the part presented in the Introduction with the addition of a size and a form tolerance applied to each hole. Each feature is represented by three nodes to represent the three tolerance zones of position, size and form associated with it.

9.4 Composite features

So far, the relationships have been explained with regard to simple features. There are two ways that composite features can be handled. The first method is to have a node in the network for each tolerance zone of the composite feature. This essentially treats a composite feature in the same way as a simple feature and assumes that tolerances

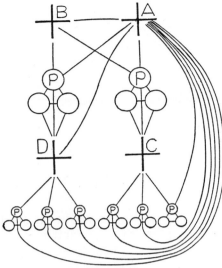

Fig. 9.18

have been applied to the composite feature as a whole. However, if a different tolerance has been applied to each simple feature then the second method must be applied: the zones of each feature must appear as separate nodes in the network. A difficulty is that there may be a datum which is the centre plane, line or point of the composite feature. The interesting problem with composite features is to find the constraints that such a datum has relative to each of the zones of the simple features. This is the problem considered here.

Any symmetric composite feature can define a datum and the position of the datum is affected by the surface of each of the component simple features. Fig. 9.19 shows the network of zones and datums for a composite feature consisting of three simple features each with form, size and position tolerances.

The constraints on the position of the datum relative to the zone can be worked out using the two stages given for simple features in in the discussion of relationships between a datum and a zone.

Stage 1: Determining the maximum size of the measuring solid

First define a **maximum material solid** (MMS) of a feature to be the union of the inside of the zone and the totally material region of the zone. Also, define the **least material solid** (LMS) to be the totally material region alone.

Consider a composite feature consisting of two parallel cylinders. In Fig. 9.20(a) the zones of size tolerance of the component simple features are shown at their nominal positions. The maximum size of the

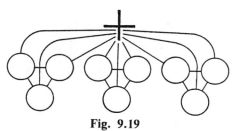

Fig. 9.19

measuring solid that is required for a real feature lying inside these zones is such that its surface is coincident with the outer surface of the zones. However, the relative position of the zones is not fixed. The effect of moving the zones away from their nominal positions is to increase the required size of the measuring solid. For instance, suppose that one of the zones is displaced translationally as in Fig. 9.20(b). It is not possible to enclose the MMS simply by translating the measuring solid. (Recall that the measuring solid is a scaled and displaced copy of the nominal feature and so the distance and angle between its two components are fixed.) Instead it is necessary to increase the size of the measuring solid. The increase is equal to half of the displacement undergone by the zone.

Also consider the case where one of the cylinders is tilted. (Fig. 9.20(c)). Again the measuring solid must be increased in size if it is to enclose the MMS. The amount of increase is half the change in angle multiplied by the length of the cylinder.

The relative location of the zones is constrained by other relationships. Knowledge of these constraints allows the maximum size of the measuring solid to be determined. The method is to apply to SUPINF algorithm described by Brooks [2] to a signed distanced expression associated with the feature. However, there is no room here to go into details.

Stage 2: The constraints on the measuring solid relative to the least material solid

Having found the maximum size for the measuring solid we can move on to find the constraint on the position of the datum relative to a zone of each of the simple features. Use is made of the fact that the LMS of the composite feature must be enclosed by the measuring solid. Therefore, the signed distance between the complement of the measuring solid and the LMS must be greater than zero. It follows that the signed distance between the complement of the measuring solid and each component of the LMS corresponding to the simple features must be greater than zero. As a result an algebraic constraint on the position of the datum relative to a zone of each simple feature is found. During

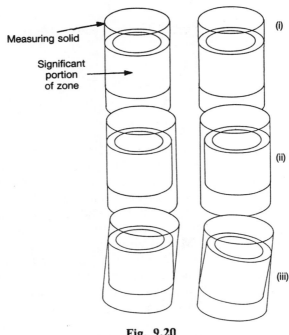

Measuring solid

Significant
portion
of zone

(i)

(ii)

(iii)

Fig. 9.20

these calculations, we must be pessimistic and assume that the measuring solid has a maximum size.

The constraints thus obtained are, as usual, only an upper bound and there is no guarantee that they are realistic. Although for some simple features it can be shown that the constraints do represent the worst possible case, in general this is not true.

9.5   Conclusion

This paper has shown how a geometrically toleranced part can be represented in a computer. The geometry of relationships between tolerance zones and datums has been analysed. Parts with unknown and variable shape are represented with datums and tolerance zones which are ideal geometric objects. The representation allows each feature of the part to have tolerances of form, size, orientation, absolute position as defined by Requicha. Although not included here, other tolerances, notably the **maximum metal condition** position tolerance, can also be represented.

The network of zones and datums has arcs with which constraints can be associated. The derivation of these constraints is made by two types of techniques. Firstly, there is the case-by-case analysis. A catalogue would be provided with constraints for all situations that can occur. In the relationships between datums in a datum-system and between a datum and a zone located with respects to it there are only a small number of possible situations and so can easily be catalogued.

Secondly, some constraints are derived using signed distance. A catalogue could be available of signed distance expressions for different shapes of feature. Signed distance expressions only have to be known for simple features: constraints on relationships involving composite features can be evaluated using signed distance expressions for their component simple features.

Constraints obtained by these methods are often approximate. However, the approximations are either small enough to be ignored or produce an upper bound of the variational class of the part. It has been shown that useful results can still be obtained under such approximations.

References

1.   British Standard 308 (Part 2: 'Dimensioning and tolerancing of size', Part 3: 'Geometric tolerancing'), 1964.

2.   R.A. Brooks, 'Symbolic reasoning among 3-D models and 2-D images', Stanford University Department of Computer Science Artificial Intelligence Memorandum 343, June 1981.

3.   A.A.G. Requicha, 'Toward a theory of geometrical tolerancing', University of Rochester Production Automation Project Technical Memorandum 40, March 1983.

Discussion

D. Williams: When you're putting assemblies together they're often mechanisms. Have you any way of handling the degrees of freedom associated with things being a mechanism? For instance, in a gearbox, everything has a rotational degree of freedom.

Fleming: I've only considered static assemblies.

D. Williams: In a static assembly, you may only be interested in interference, but if anything has slop in it, then the assembly is likely to be a mechanism.

Fleming: If slop is negative, then you have interference! But I've given no thought to representing mechanisms, except in static situations.

Middleditch: You made a point about cause and effect. It seems to me that, once you've generated these algebraic constraints, they can be interpreted either way around. So you can work backwards: cause and effect are not so important; you have a sequence of events which can run both ways. Is that true?

Fleming: Although a network can be interpreted in either direction, it helps, when you come to analyse it, to think how you actually arrived at that particular network. There are paths through it, but when you look back to the actual geometry, you see that they don't have any actual meaning.

Middleditch: Is that because you're going backwards, or is it because you're both backwards and forwards. When (Fig. 9.5) you went down to C and back up, you went backwards and forwards. If you went consistently backwards you'd be all right.

Fleming: It would also be OK to go up and then down, because of the way in which the constraints are related.

Parslow: It seems to me that, if you have a complicated thing, you're going to have cumulative tolerances building up. It's going to be meaningless after a while.

Jared: Nothing will fit.

Parslow: Yes. If you construct your tolerances each on the last toleranced feature, you're in the position of someone doing a survey. What they do in surveys is constantly to refer back and triangulate, and reset the basis all the time. It looks to me as if you're starting off and going down; you're going to have cumulative errors building up, which are going to get out of hand after a while.

D. Williams: That depends dramatically on the style of dimension that you put on the part. You don't dimension like that in practice.

Parslow: That's what I'm saying.

Todd: You can do what a surveyor does, and refer everything back to your original datum.

D. Williams: Can your system cope with redundant dimensions? Engineers occasionally throw in redundant dimensions: perhaps to make the drawing easier to read. They're extra constraints that don't actually fit into the system.

Fleming: I suppose that, if they were specified as being nominal dimensions, then they would be ignored anyway.

D. Williams: You must also be able to test the correctness of dimensioning.

Fleming: If a dimension wasn't included in the network, then you could then go back and work out what it actually was. I haven't thought about what would happen of you over-dimensioned by mistake, so that there was too much in the network. It would be difficult to detect that; I'm not quite sure what would happen.

Sabin: There is a theorem which D.B. Welbourn was fond of quoting, which is that every drawing is over-dimensioned *and* under-dimensioned, and wrong

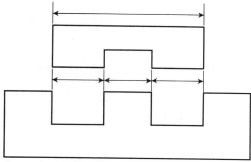

Fig. 9.21   Each dimension arrow indicates a potential fouling, which therefore needs tolerancing, although all four nominal dimensions can be deduced from any three of them.

somewhere else as well! It is typical that the dimensioning will not in fact be kinematically determinate. The result, as Hillyard found out [5], is that there are a lot of implicit dimensions: the two sides of a thing which look rectangular actually *are* rectangular unless you explicitly dimension something else. Two things which look parallel are intended to be are, unless you actually have some evidence that they are not. There is implicit dimensioning there. My third point is that, if you assemble an E shape with a U shape (Fig. 9.21), so that they are interdigitating, there actually three dimensions that you need to specify with this configuration, but *four* tolerances, to avoid clash. You may have to over-dimension, in order to have enough dimensions to hang the tolerance on.

Murray: How easy would it be from your scheme actually to draw the maximum metal condition and the minimum metal condition for the part?

Fleming: I haven't considered drawing at all.

Parslow: I'm still worried about the order in which you're doing things. It's very specific (Figs. 9.3 and 9.4). Suppose one of the large circles doesn't happen to be central; it has moved sideways. As I see it, the tolerancing indicates that you think that you'll be able to get the dial down on to the studs. However, if the hole hasn't been cut properly, so you've moved the datum over, you're going to have two parts made up which won't fit.

Fleming: That is just the sort of thing that I want be able to predict.

Parslow: You can't do it in this case, because they're two different parts.

Fleming: To deal with that I would have to think of it as an assembly, and create the network for both parts joined together with relationships representing contacts.

Parslow: I'm just under the impression that you'd organized the tolerances in such a way that, if it got out of tolerance, it would still appear able to fit. It looks to me as though you're taking allowance of the tolerance in moving your datums out because the thing isn't cut exactly as you expected.

Fleming: Perhaps I hadn't explained. What I'm doing is, given a tolerance specification, asking whether parts will actually be functional.

D. Williams: I think maybe that pitch-circle problem is rather a special one, which arises out of very traditional manufacturing methods, where you used a jig which is centred on the large hole, whereas now you might use an NC machine tool to generate the holes, all from a single datum off the machine tool. Still, it's the traditional way to dimension a pitch circle.

Cameron: You've said that you're converting your constraints into an algebraic form. What sort of tools would you be using after that?

Fleming: I'm using Super-Nymph; it's similar to the system mentioned in Fisher and Orr's paper [4].

Cameron: Does that appear to be adequate?

Fleming: Yes, I think so. The constraints would be linear, because you can make approximations; the uncertainties are very small. Although zones may be inclined relative to one another, it's by a very very small amount.

Cameron: You're taking a first-order approximation.

Wright: Having got the ability to reason about the tolerances on a drawing, have you given some thought as how that might be used to interface back to a designer? If he was sitting at his CAD database, specifying an assembly of two parts that he wanted to fit together, you might be able to reason about his dimensioning and provide him with information. For instance, he could be informed that he hadn't provided enough information. Alternatively, he might be warned that certain tolerances could accumulate and give a problem, such that the assembly might not fit together.

Fleming: What could be done is to put variables on given tolerances, so that the constraints have extra variables in them: the result has extra variables, in which you can put constraints, and tell the designer a dimension has to be within certain bounds. A problem could be that you'd end up with too many variables. The designer could suggest what things he wants to vary.

Wright: Have you been thinking about it as an interactive tool?

Fleming: I haven't been considering that, no. The application is the RAPT robot language. The initial motivation for the work was dealing with uncertainties in assembly by robots. You know what tolerances are on the parts and want to put the parts together.

Additional references

4.   R.B. Fisher and M.J.L. Orr, 'Geometric constraints from 2½D sketch data and object models', in *Geometric Reasoning* (Proceedings of a conference held at the IBM UK Scientific Centre, December 1986), OUP, 1989.

5.   R.C. Hilliard and I.C. Braid, 'Characterizing non-ideal shapes in terms of dimensions and tolerances', *Computer Graphics* 12,3 (Proceedings of the ACM SIGGRAPH 78 Conference, Atlanta, Georgia) (234-238), 1978.

10 Recognizing and using geometric features

G. E. M. JARED

10.1 Introduction

This paper is primarily concerned with the automatic recognition of the geometric features of an object from its description held in a computer as a solid model. It may perhaps be surprising, in view of the much-vaunted claims for solid models as 'complete geometric descriptions of three-dimensional objects', that is necessary to discuss additions to such systems for the extraction of feature information. However, as we shall see in the brief résumé of solid model representations that follows this Introduction, solid modellers were designed for certain specific purposes and do not *explicitly* hold high-level information about the 'structure' of objects. Thus, in order to obtain information about the geometric features of an object, extra information has either to be 'grafted on' to the solid model or, as in the case of automatic recognition, inferred from existing data structures.

It is appropriate at this point to consider what a feature is. Pratt and Wilson [10] define it as a region of interest on the surface of a part. This apparently vague, all-embracing definition probably results from the wide divergence of opinion of practitioners in the many fields of mechanical engineering design and manufacture as to what aspects of a part constitute features. This paper will concentrate on what, for want of a better term, might be called **geometric features** which are largely depressions in, or protrusions from, the surface of an object. However, even this definition is not as tight as it perhaps should be since it is not easy to define the range of features that fall into this category. Furthermore, there certainly is no large measure of agreement as to the types of feature that make up the set of geometric features. We have only to consider the old story, illustrating the difference between design conception and the manufacturing processes used, of the designer who sees a part as a formed piece of metal and the production engineer who sees the same part as a collection of holes that have to be made which are almost incidentally surrounded by material. Thus we find that not only is there a vast range of entities that someone in an organization will consider to be a feature, but also that there may well be many ways of describing the same object in terms of different types of feature. The only safe conclusion is that there is unlikely to be a unique set of features for any one part.

When considering strategies to allow the manipulation of the geometric features of a part two—probably complementary—options present themselves. The first is to store feature information in data structures that are linked to, or even integrated, with the solid model. The second is to infer feature information from the existing solid model by what amounts to the recognition of patterns within the data. Implementation of the first option might allow the user of a solid modelling system to define an object by successively adding (or subtracting) features. The use of the second option might allow the user to define objects in a slightly less constrained manner with an *a posteriori* step of recognizing the features of the object that has been created, whether this is by automatic recognition or by some degree of 'interactive feature recognition'.

An alternative to the first approach of direct definition of features is exemplified by the Alvey 'Design to Product' project [13] particularly the Edinburgh Designer System [9]. In this system the master definition of a product is held in terms of facts about its function, required performance and so forth, and the corresponding solid model is no more than a particular geometric instantiation of it. It might be imagined that such an approach removes the need for feature recognition, since it has often been suggested as a means of inferring information from the solid model which in this system would be explicitly held in the master product definition. This is arguably the case, but if we accept that much of the wide diversity of feature definitions comes from the tasks downstream from design, then it is quite likely that several completely new sets of features for the same part will be needed for different applications. It is here that feature recognition techniques could play a role.

In the remainder of this paper, the various types of solid modeller will be very briefly examined in order to illustrate the low-level nature of the descriptions of solids that they use; then some possible and actual approaches to the explicit representation of features will be discussed; next, research on automatic feature recognition at the University of Cambridge and Cranfield Institute of Technology will be described; finally, some concluding remarks and speculations on future directions in solid modelling and feature recognition will be made.

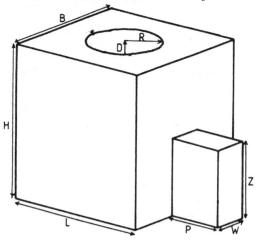

WIDGET (H, L, B, D, R, P, W, Z)

Fig. 10.1 A solid model represented by pure primitive instancing.

10.2 Types of solid modeller

The following is not intended to be an exhaustive and rigorous survey of the various types of solid modeller (for this the reader is referred to articles such as that by Requicha [11]), but a brief description of each approach with an emphasis on its ability to cope with the representation and manipulation of geometric features.

Several authors, including Miller [8], have attempted to set down criteria for comparison between solid modelling representations; amongst them is the concept of so-called **uniqueness**. This can be loosely defined as measuring whether there is a direct correspondence between an object and its model or whether there are several different ways of representing the same object. Clearly this property has some relevance to object recognition. On a global scale, if an object has a unique representation then identification and comparison of models to determine whether they correspond to the same object is made simpler. It also seems intuitively likely that uniqueness will govern the ease with which certain patterns within a model can be associated with particular geometric features in the object.

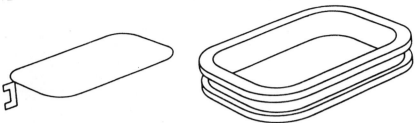

Fig. 10.2 A solid model represented by generalized sweeping.

Pure primitive instancing

In this type of modeller an object is represented as an occurrence or instance of a particular parameterized template for a class of objects (as illustrated in Fig. 10.1). Primitive instancing is not widely used because of the limited domain of objects that can be represented unless an impractically large range of templates are provided, although its use does confer certain important advantages. An object will have a unique representation in terms of a particular template and an associated set of parameters. Thus, in a parts database for instance, it will be easy to distinguish between parts and to identify a part from the template used to create it. Similarly, there is no need for feature recognition since the features of a family of parts could be arranged to be an inherent part of their template.

Generalized sweeping

This type of solid model is one of the least well understood and this has led to a lack of implementations in practical systems. Objects are described as the name suggests by sweeping either of a profile curve or a solid along another curve in three-dimensional space (as illustrated in Fig. 10.2). The lack of understanding of this type of model results from the fact that, even for profiles and paths made up of curves with relatively simple geometry, the surfaces that result can be mathematically complex and intractable. Unfortunately, it seems intuitively likely that in such models there would be a strong correlation between the geometric features of an object and the structure of the corresponding model—certainly for prismatic and rotational parts.

Cellular decomposition

The most general form of a cellular decomposition model is of arbitrarily shaped 'bricks' juxtaposed in three-dimensional space (as illustrated in

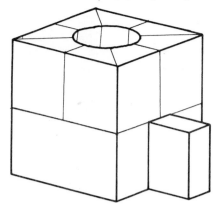

Fig. 10.3 A solid model represented by cellular decomposition.

Fig. 10.3). The bricks or cells may be any polyhedron, may have curvilinear faces, but may not themselves contain voids. The range of parts that could be described by a cellular decomposition system is potentially large and depends on the available types of surface geometry of the cells and the variation in cell sizes allowed. If the cell polyhedra chosen are simple shapes whose properties, such as volume, may be easily calculated, then it becomes equally easy to calculate the properties of the complete solid; this is one of the advantages of cellular decomposition models.

It is clearly unlikely that there will be, in general, a unique arrangement of cells to represent any one object, which makes it more difficult to represent and manipulate geometric features. It will not be easy to compare two objects to determine whether they contain the same feature purely by examining the arrangement of cells. It will be equally difficult to recognize features in an object by searching for particular patterns of cells. This applies equally to general cellular decomposition models, as just described, and to the two special cases that follow.

Although it is not necessary for the cells of this type of model to be identical or to be regular polyhedra or be arranged on a regular grid, there are some advantages in terms of simplification of algorithms and data structures is they are. One example of such a simplification is so-called **spatial occupancy enumeration**. In this type of model the region of space where an object lies is divided up into a regular, rectangular grid of cells and those cells which are wholly or partially occupied by it are marked or **enumerated** (as illustrated in Fig. 10.4). The major difficulty with this type of model is that it is an approximation to the object and is only as accurate as the cell size allows, furthermore, a huge number of cells have to be stored to give reasonable geometric resolution. However, with the recent dramatic increases in the memory capacities of computers and the possibilities for devising special

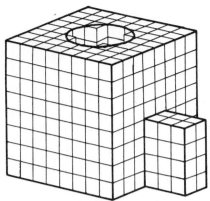

Fig. 10.4 A solid model represented by spatial occupancy enumeration.

hardware that could deal with many cells in parallel, this type of model may yet become more widely used.

In a spatial enumeration model it is quite likely that large numbers of cells will be taken up in representing the interior of an object where there is little change taking place—all the cells are full—just so that a sufficiently small cell size is used to give a good approximation to the object's surface. The so-called **oct-tree** model aims to alleviate this problem by using a variable resolution with large numbers of small cells being used only where necessary. The model may be seen as a series of layers, in the first of which the object is approximated by a cube. In the second and subsequent layers cubes in the previous layer are subdivided into eight sub-cubes or octants which are marked as being full, empty or partially filled by the object. It is only necessary to sub-divide partially filled octants on succeeding layers in order to gain a more accurate representation of the object. This type of model is illustrated in Fig. 10.5.

Both spatial occupancy enumeration and oct-tree models suffer from the same drawback of being approximate representations of a solid. They will use cuboidal cells of a finite size to represent an object's boundary and thus have inherently to approximate both curved and sloping planar faces. This seems likely to be a major disadvantage in representing or recognizing geometric features which are often strongly linked to surface description.

Set-theoretic models

The set-theoretic model of a solid is built up by combining together simple elements using so-called **Boolean operations**. This can be applied to the construction of complex solids from simple ones (called **primitives**) as illustrated in Fig. 10.6, or to the modelling of primitives

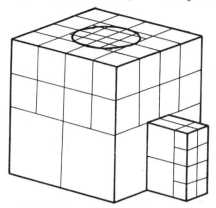

Fig. 10.5 A solid model represented by oct-tree decomposition.

by the combination of more low-level entities known as **half-spaces**. A half-space, as its name might imply, is a surface that completely divides space into two portions—one each 'side' of the surface. Thus an infinite plane may be a half-space which divides space into that on the left and that on the right of it and a spherical surface divides space into the region outside the sphere and that inside it. Boolean operations are defined as operating on sets of points in three-dimensional space and producing combinations of those sets as their result, the operations available in most set-theoretic modellers are **union**, **intersection** and **difference** (which last may be expressed by a combination of negation and intersection).

There is clearly unlikely to be a unique way of combining solids or half-spaces to produce a description of a particular object; it is only necessary to consider the case of adding two mutually annihilating, redundant primitives to an existing object description to prove this. This lack of 'uniqueness' of models means that objects cannot be established as identical merely by comparing their representation by primitives and operations, and also that is not possible to discover the presence of geometric features by searching for particular patterns in the model. There is a further difficulty in that one primitive may contribute to more than one geometric feature.

Boundary representation

As its name suggests, this type of model represents a solid by storing a description of its boundary. This description is most often a so-called **face-edge-vertex graph** which shows the way in which the regions of an object's boundary (faces) are arranged with the edges which bound those regions and are connections between them at vertices (as illustrated in Fig. 10.7). The network has associated with it the

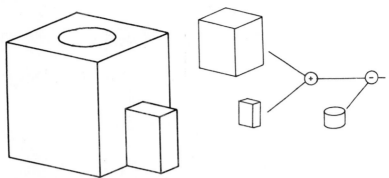

Fig. 10.6 The set-theoretic representation of a solid model.

appropriate geometric information for each entity: surfaces with faces, curves with edges and coordinates with vertices. Practical data structures for boundary representation modellers tend to have a large degree of redundancy in the pointers used to make associations between elements of the face-edge-vertex graph in order to reduce the amount of searching necessary to traverse the structure. This means that boundary representation models tend to occupy larger amounts of store than, for example, set-theoretic ones, but this is to some extent compensated by the large amount of explicit information they contain which considerably speeds up interrogation of the model.

10.3 Feature representation

Exploration of the representation of geometric features in solid models has, as far as can be ascertained, so far been limited to research on the set-theoretic and boundary representation types. One of the larger studies of the topic is the volume of conceptual work that has been carried out for the CAM-I organization [10] which is limited to consideration of these two representations.

As previously stated, geometric features may be represented either by data structures that are an integral part of the solid model or by external or secondary models that have some link or association with entities within the solid model. An example of the latter technique is seen in some recent versions of the ROMULUS modeller [12], where it is possible for the user interactively to declare associations of faces together into so-called **facesets**. It is then possible to operate on such facesets as single entities. This facility can serve several purposes—not only the representation of geometric features—since there are no specific constraints on the relationships between faces that can be grouped together in this manner. The disadvantage of using such groupings

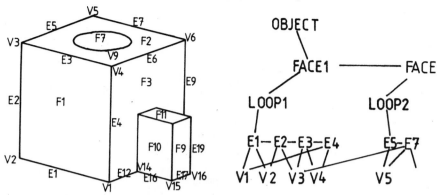

Fig. 10.7 The boundary representation of a solid model.

external to the solid model to represent features is that it is not easy to maintain consistency between the two structures. Faces may disappear from or be created in the solid model as a result of modifications and the faceset model has somehow to be updated in step with it.

An example of the representation of features by data structures integrated with the solid model is illustrated by the work of Hailstone [4]. He describes a simple solid modeller in which the user can interactively tag faces as belonging to a certain type of feature, chosen from a limited selection built in to the system. The integration of the feature representation and the solid model is exploited to aid user input of feature descriptions and to perform a certain level of checking. The input of features may be performed by interactively pointing at faces on the screen and indicating that they are 'characteristic' faces of a particular feature type—for example, the base of a pocket or the top of a boss. It is then possible for the system check the surrounding faces to ensure that such a feature could exist at that place and in doing so to collect together the appropriate faces. In addition to simple geometric features, so-called **compound features** may be declared by interactively associating groups of features (Fig. 10.8).

There are certain attractions in providing interfaces to solid modellers which only allow the user to construct objects from a given range of features. The major advantage is that design codes or other constraints may be enforced by limiting the range of features available and the operations that may be used on them. The major drawback is that, unless some complex function analogous to feature recognition is provided, this type of system enforces a unique view of the features of a part and thus is likely to be restricted to only a few applications that subscribe to that view. This latter point appears to be borne out in practice, in that systems implemented thus far which allow design by features have been limited to particular applications such as generation of NC data for milling.

Geometric Reasoning

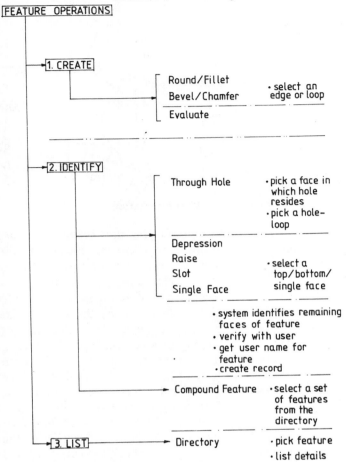

Fig. 10.8 The general arrangement of operations for interactive feature input.

10.4 Automatic feature recognition

As far as is known, with the exception of Woo [16], most work on feature recognition from solid models has been carried out using boundary representations, although this is not to say that other representations cannot be used.

In a paper given in 1982 [16] Woo describes a method of decomposing a boundary representation solid model into a Boolean combination of solids as would be used in a set-theoretic model. The set-theoretic tree resulting from this process is made up of alternating additions and subtractions of convex components. This representation, which Woo calls an **alternating sum of volumes**, is built up by a process of taking convex hulls and comparison. Although the process produces a unique decomposition of objects into convex components, the algorithm unfortunately does not converge in all cases.

Whereas Woo was concerned with the decomposition of solids into simpler sub-volumes other workers have developed algorithms and techniques for finding particular configurations in the face-edge-vertex graph representing an object's boundary. Many projects on feature recognition are concerned with one or a small number of specific applications; this is perhaps not surprising in view of the earlier statement that there is unlikely to be a unique set of features in a part that is universal to all applications. An example of this is research at Purdue University [6,7] on using syntactic pattern recognition to discover strings of elements in a part description that correspond to features that are significant in operations planning for machining rotational parts.

Other feature recognizers use a grammar-based approach with techniques similar to those used in two dimensions and spatial problems in other application areas [6]. **Shape grammars** [2,14] are an example of such techniques; they were originally 'discovered' and used in the field of architecture as a means of formally describing the style and form of buildings. Thus, for example, a grammar was devised [15] which could generate patterns of rooms for villas in the style of Palladio. The initial applications of shape grammar techniques were 'generative', that is to say they rules were set up and used to generate patterns. In the application of similar techniques to automatic feature recognition that will be described in the remainder of this section, a grammar is used as a tool for 'parsing' a solid: forming a description of how the basic shape elements contained in a grammar are juxtaposed in order to make up that solid.

As with other investigations of automatic feature recognition, the work of Kyprianou at Cambridge [5] was concerned with a particular application, in that case the classification of mechanical engineering parts and the generation of part codes. The research resulted in the implementation of a program that could take the solid model of a part and automatically produce a code for it in a standard, manual classification scheme. The program was intended to cope with different classification schemes and had a meta-language input to describe an individual scheme. As a result of this independence of specific schemes a means had to be found of decomposing solids into basic elements that could be compatible with every scheme. Thus feature recognition took place in two phases; an initial partitioning into low-level features and then a further collection into higher-level entities.

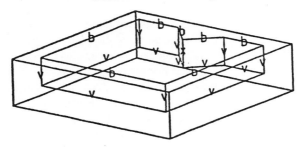

v concave edges in the depression
b edges forming the convex boundary
x convex edge shared by faces of the depression

Fig. 10.9 A depression with convex edges inside its boundary which can be recognized as a feature. The edges are marked as follows: v, concave edges in the depression; b, edges forming the concave boundary; x, concave edges shaded by faces of the depression.

In his thesis, Kyprianou creates a new type of grammar, related to shape grammars, known as a **feature grammar**. The basic elements of a feature grammar are entities which occur in the face-edge-vertex graph of a boundary representation solid model, such as convex, concave, or smooth edges. (A smooth edge is one where there is no change in surface normal between the faces either side of it.) The information as to whether an edge is concave, convex, or smooth is not maintained by the modeller used by Kyprianou and so it has to be obtained in a pre-processing phase of the feature recognizer by a process of visiting all edges, examining their adjacent faces and marking them appropriately. This pre-processor also accumulates information concerning the number and type of edges that are connected to each vertex of the model and the type of edges that make up every loop of edges in the model. Then, in the main phase of recognition, rules from the feature grammar are used to recognize collections of these basic elements as geometric features. For example, a loop of convex edges surrounding a network of concave edges will form a depression such as a pocket, although some account has to be taken of special cases such as shown in Fig. 10.9.

The recognizer described by Kyprianou is based on feature grammars, but has certain extra techniques for speeding up the recognition process. The faces of the solid model are analysed to determine what is called their **morphological** content which can be roughly translated as the likelihood that they border or contain some feature. The recognizer looks for **primary faces** (Fig. 10.10) such as those with inner loops or bordering a concave edge or which have a concave surface. The primary faces are used as start points for scanning the edge network in order to divide up the model into sets of faces

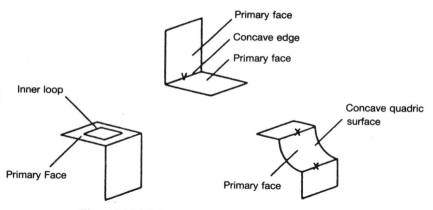

Fig. 10.10 Primary faces in feature recognition.

corresponding to depressions or protrusions on its surface. Further investigation of the edge networks within such broadly classified features can then allow more accurate categorization of, for example, depressions as pockets or slots. Thus the basic structure of feature description produced by this recognizer is determined almost entirely from consideration of the edge network and is also governed by the positions of primary faces. As previously stated, the more complex relationships used for producing part codes are determined in a second pass which checks the geometric information attached to the face-edge-vertex graph in order to discover, for example, sets of faces lying on cylindrical surfaces whose axes are coincident and thus determine whether a part may be classified as rotational.

In his thesis, Kyprianou lists several problems encountered with his feature recognition techniques such as difficulties in identifying 'nested depressions' or 'implicit protrusions' (as illustrated in Fig. 10.11). These and other problems may be ascribed to deficiencies in the solid model or, rather more correctly, to the absence of data structures in the model representing relationships that are needed for recognition. The rules used by the recognizer are really about relationships between faces and these are not necessarily expressed explicitly in the edges and loops of the solid model. Thus, for example, some protrusions can be recognized because they spring from a single loop of concave edges in one face, whereas so-called implicit protrusions spring from a set of concave edges that are connected but belong to different loops and faces. A further problem for recognition imposed by the solid model representation is that of so-called **fake edges**. These are edges arbitrarily inserted into curved surfaces in order to divide them into areas which may be uniquely projected onto a plane. This is done so that interrogations of curved faces may be implemented more simply as calculations on its projection on a plane.

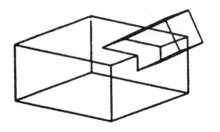 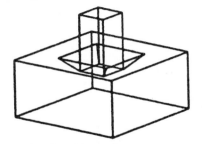

Fig. 10.11 Implicit protrusions, which do not arise from inner loops, are one of the problems in feature recognition.

Although the feature grammar used by Kyprianou is outlined in his thesis, it is not easy to assess the effect of changes to the grammar and to the associated solid modeller in trying to overcome the recognition problems described above. This is because the feature grammar is 'hard-wired' into the recognizer as are the apparently arbitrary criteria for choosing so-called primary faces. It would be desirable to have a flexible recognizer that could accept different feature grammars and allow different parsing strategies. This objective, together with modifications to the solid model to assist feature recognition and representation, is the subject of continuing research at Cranfield [3].

10.5 Concluding remarks

It may seem attractive to reject feature recognition as a difficult and perhaps impossible task, in favour of concentrating on representation of features and user input of feature information. However, in the author's opinion, to do so would be to lose some potentially useful tools for 'geometric reasoning'. It may not be possible to devise recognizers that have a one hundred per cent success rate and produce a list of features that correspond to a user's intuitive view of a part, but it certainly seems likely that recognition techniques can be used to answer a different question. If instead of asking for global recognition—'what is this object?'—the question is posed in terms of 'does this object contain features like this?' then a successful outcome is more likely. Such interrogations of a solid model can still provide the basis for the automation of 'low-level tasks' in design and manufacturing. They could, for example, provide support for simple rule-based tools to advise designers on the implications of their present course for downstream activities in production of which they have little knowledge or experience.

A point of concern about solid modelling that emerges from this consideration of feature representation and recognition is that modellers are still complex, ill-understood monsters that can only be handled by a few skilled 'magicians'. Modellers should simply be tools for geometric representation that are taken for granted in a similar manner as compilers are for programming; at present *they are not!* To some extent this results from a lack of vision and direction by modeller developers. If modellers, which tend by their very nature to be large programs, are not created using tools with a high degree of abstraction then they are unlikely to be flexible enough to be used as tools themselves. To say that 'FORTRAN is the *de facto* standard for CAD/CAM programs' is to confuse issues of portability and widening of applications with the ability to write in a familiar language. Mixed-language programming environments and standardization of software and hardware interfaces offer a much more sensible path than fixing on a particular programming language.

A further point to be considered is that, whilst manufactured objects tend to be solid, the engineering idealizations of them used in design and production simulation may well be one or two-dimensional: for example, wire-frame, sheet or cellular constructions. It is also possible that all of these types of idealization may be used in different parts of the same product. Thus far we have only considered feature recognition in the context of solid models. In the future it might well be profitable to examine techniques for the automatic construction of idealized models and for the 'intelligent extraction' of data from them.

References

1. B.K. Choi, M.M. Barash and D.C. Anderson, 'Automatic recognition of machined surfaces from a three-dimensional solid model', *CAD Journal* 16, 2, 1984.

2. C.F. Earl, 'Shape grammars and the generation of designs', in *Principles of Computer Aided Design* (J. Rooney and P. Steadman, eds), Pitman and the Open University, London, 1987.

3. G.E.M. Jared, 'AI Europa paper', Proceedings of the AI EUROPE 86 International Conference, Wiesbaden, TCM Expositions and Liphook, September 1986.

4. S.R. Hailstone, 'Explicit Form Feature in Solid Modelling', MSc Thesis, Cranfield Institute of Technology, 1985.

5. L.K. Kyprianou, 'Shape classification in computer aided design', PhD Thesis, University of Cambridge, 1980.

6. R.G. Lauzzana, 'A field grammar formulation of the image recognition problem', Proceedings of the AI EUROPE 86 International Conference, Wiesbaden, TCM Expositions and Liphook, September 1986.

7. C.R. Liu and R. Srinivasan, 'Generative process planning using syntactic pattern recognition', *Computers in Mechanical Engineering*, March 1984.

8. J.R. Miller, 'Introduction to solid modeling', Tutorial No. 18, ACM SIGGRAPH Conference, San Francisco, July 1985.

9. R.J. Popplestone, 'The application of artificial intelligence techniques to design systems', Proceedings of the Japan Society for Precision Engineering International Symposium on Design and Synthesis, Tokyo, 1984.

10. M.J. Pratt and P.R. Wilson, 'Requirements for support of form features in a solid modeling system', CAM-I Report R-85-ASPP-01, 1985.

11. A.A.G. Requicha, 'Representations of rigid solids; theory methods and systems', *ACM Computer Surveys* **12**,4 (437-464), 1980.

12. Shape Data Ltd., 'Benchmarking results—ROMULUS', Proceedings of the 3rd CAM-I Geometric Modelling Seminar, CAM-I Document P-85-MM-01, 1985.

13. T. Smithers, 'The Alvey large scale demonstrator project "Design to Product"', Proceedings of the 3rd International Conference on Advanced Information Technology, Zurich, November 1985.

14. G. Stiny, 'Pictorial and formal aspects of Shape Grammars', Birkhause, Basel, 1975.

15. G. Stiny and W.J. Mitchell, 'The Palladian Grammar', *Environment and Planning B* **5**, 1987.

16. T.C. Woo, 'Feature extraction by volume decomposition', Proceedings of an MIT Conference on CAD/CAM Technology in Mechanical Engineering, 1982.

Discussion

Geisow: Just to bring things a full circle, I think that Gips, who worked with Stiny, actually used shape grammars to recognize three-dimensional views of objects [18].

Jared: Unfortunately from our point of view, he was concerned with getting three-dimensional information from two-dimensional graphics, as I recollect. He didn't have a solid modeller which he could have used to try to recognize features.

Martin: Given that you started off by saying that you thought that features in some sense provided depressions or protuberances of an object, wouldn't one method of feature recognition be to look for complicated faces, in the sense of faces with not just an outside boundary, but also some kind of inside boundary. Maybe do something like taking plane sections above and below that face, to see whether something is sticking out or sticking in.

Jared: Yes, but as you say, Kyprianou picked up on that using the topological side of things: using the fact that there was an internal loop, without looking at the geometry. I suspect that his motivation in doing that was that it's relatively cheap. His recognition runs very quickly indeed; it may not tell you very much, but it's interactive! He was working on part coding classification, and started to look at geometry in a later phase. Having recognized gross structure, he then started looking at things like cylindrical surfaces to see whether they were aligned or not, so that he might find a complete alignment of holes through a part; something like that. When you start doing that, the whole operation starts to get rather more expensive, because you have to do geometric interrogations, rather than just using simple Boolean facts about how things are connected together. As you probably know, that's one of the more expensive things that you can do in a geometric modeller. It would be nice to be able to continue to look at reasonably simple concepts, and to be able to refine them.

Sabin: How long would it take you to detect the fact that a particular collection of cylinders was a counterbored hole—to have a slightly more complicated thing to recognize?

Jared: You would unfortunately not find that in the first place; the technique carefully partitioned the thing up into two bits. The counterbore is going to give you a primary face, and so it is not going to recognize the other bit of it. So that must be a process on top of the initial parsing of this lexical network: to go back and discover that, of these two pockets, one of which is a leaf-based end from that branch, they do happen to have the same axis, and it does rather seem as though this might be a counterbore. But is another layer over the top, if you like.

As I understand it, Kyprianou embedded that sort of consideration in a rather nasty incomprehensible meta-language which was part of the part-coding piece of what he did. He basically partitioned things up into nice small compartments at the feature recognition stage, and then, in his actual part coding, he started to put things together into higher-level entities. This is the sort of thing I talked about at the beginning; why not look at basic feature elements from which you can build up other things? His meta-language for describing how you put things together—I wouldn't insult it by saying that it's incomprehensible—I'd just say that I'm too stupid to understand it.

Forrest: The data structure you are using in BUILD is a winged-edge data structure?

Jared: Heavily modified, yes.

Forrest: Most of the features that you are talking about are essentially face features. Now there was a paper in Siggraph 85 [17] about a face-based boundary representation. Would that have made things much easier? In retrospect, if you were to build another modeller, would you build a face-based, rather than an edge-based system?

Jared: Maybe. I'd like to tell you after I'd done both, rather than speculate. To take your question a little more seriously, the things that you're actually expressing are relationships between faces, so that might be quite a profitable approach to look at. But it has taken rather a long time to generate the modeller I've got, so, unless I come to a complete full-stop with this, I'm not prepared to indulge in a complete re-casting: though it might be interesting to try some experiments in that way.

This brings me to some extra remarks I would like to make, which might stir up some controversy. We've done what we've done in order to try to improve solid modellers, and, as I said at the beginning of my talk, modellers were supposed to be the bee's knees: the answer to everything. it was also proposed, in the early days of solid modelling, that they would be rather like compilers—the sort of thing that a research student wrote for his Ph.D. thesis. Now everybody *uses* one, and they take it for granted. We've only got one honourable exception to that—Braid—who's actually gone and written a solid modeller for his Ph.D. thesis.

These tools should be much much easier to use and to manipulate. At the moment it's a black art due, in no small part to the tools which we've used to construct them; and that comes back to Forrest's question It would be nice to have a sufficiently flexible set of software tools, geometric tools, and the other tools which go into modeller construction, so that I would be able to say 'What a good idea, I'll go away and try it next week.' We're a long way away from that position. Cynically, you might see it as an argument for asking why we are doing solid modelling in the first place, but people do seem to like using some of the results.

Dodsworth: There seem to be two really quite hard problems here. One is, when you've got a lot of symmetry or pattern which you want to recognize, the other one is the interaction of features, which you've already mentioned. In one of the examples you showed, you mentioned that you had a through hole in the object. Was that geometrically a through hole, or topologically a through hole? In other words, could the top have been closed off geometrically, but the thing still think that it was a through hole? If you're going to use the model for process planning, then you are going to drill the hole and you will need to know about access. So there are a lot of geometric questions in there as well.

Jared: Yes. Those geometric questions are not tackled explicitly in the feature recognizer. It will find that it's topologically a through hole, if it's not blanked off, and if it is blanked off then it will think that's a pocket, because there won't be any connection in the topology. A through hole has the property that, if you trace in an appropriate way, from one face to its neighbouring faces, then you arrive back out on the convex boundary of the part. Whereas, with a blocked-off hole, you just go down it, and you don't come out again.

Dodsworth: Yes, but it may be blocked off such that you can't get a drill at the right diameter, but not blocked off in the sense that you could get a smaller drill through. A hole with sloping sides perhaps—designers are strange creatures!

Jared: Yes. You'd find that it was a hole, or a pocket, rather, and you'd then have to look at its geometry to find that it had sides that were sloping in. The original shape grammars and Kyprianou's work on feature recognition is all topology and connectivity, with an extra layer of geometry on top.

Effectively, rules about topology and rules about geometry are explicitly expressed. Really, what we're looking for at the moment is some sort of formalism to try to write them down and see what happens if you change them.

Martin: Surely you must have included some geometry to make certain types of decision. For instance, supposing that you've got a small cube on top of a big cube. How do you decide that the small cube is a boss on the big cube, or is the big cube a boss on the small cube? Well, perhaps that one can be decided by topology, but I think that you see what I'm getting at.

Jared: You tend to do it by ordering of loops. The thing that has to come off the inner loop is sticking out of the other bit, and you get a concave relationship with that inner loop.

Martin: That was a poorly chosen example; here's a better one. You may decide that things like tapped holes may be features in some sense, but are irrelevant to what you're trying to decide, and it is the size of them that is important: something like that.

Jared: That really comes back again to representation; do you represent your screw thread as helical surfaces, or do you have an incidental note stating that that hole is tapped. This is the implicit versus explicit feature representation problem again, which is fudged at the moment.

Martin: An even better example of what I was trying to say the first time is an L-shaped bar. You can view that either as a long bar with a little boss sticking up at its end, or as a short bar with a big boss sticking up at the end.

Todd: Or as a large block with a chunk taken out!

Jared: Indeed so: this returns us to the realization that there is no unique view of features. This present recognizer is 'hard wired'. It decides what it decides, and it's not always obvious why a particular decision was taken, and you can't easily change it. We need to be able to write down and express a formalism, which ends up with us finding out that it's a long bar with a bit sticking out, and then find out how we decided that.

Martin: But you do need to consider the geometry to decide that?

Jared: Yes. Again, the geometry that we do is very simple; it's mainly comparison. You're talking about properties which we still don't store in the geometric model. For instance, saying that a particular feature has quite a large volume or surface area. That has to be calculated. And, as I said at the beginning, this whole thing is bringing in lots of questions which solid modellers can't answer, because they haven't got the information there.

Sabin: At the risk of being pedantic, I think that at least two people at this conference wrote geometric modellers as part of their Ph.D. work.

Jared: I apologize to the other person!

Additional references

17.   S. Ansaldi, L. de Floriani and B. Falciendo, 'Geometric modeling of solid objects by using a face adjacency graph representation', *Computer Graphics* **19**,3 (Proceedings of SIGGRAPH 85 Conference) (131-140), 1985.

18.   J. Gips, *Shape Grammars and their Uses*, Birkhauser, Basel, 1975.

11 Planning automatic assembly from a CAD knowledge base

F. FALLSIDE, E. APPLETON, R. J. RICHARDS, R. C. B. SPEED
and S. WRIGHT

11.1 Introduction

This project is concerned with the development of a semi-autonomous robotic assembly system which will carry out the assembly of parts from random initial conditions. It makes use of measurements of the workspace from vision and force sensors and knowledge of parts and assembly in a knowledge base. It is aimed at small batch and varied types of assembly tasks. The project is at an early stage and this paper describes the separate modules involved, together with early results from some of them. The basis of the overall system is described first and this is followed by a description of the separate modules such as assembly planning, assembly implementation and vision systems. The work is centred round an IBM 7565 robot and will make use of CATIA and other CAD systems in its knowledge base. The project makes new demands on CAD systems by requiring them to be used on-line as part of the assembly system, for example by providing a reference three-dimensional model and simulation of the workspace which can be integrated with part recognition, assembly planning and assembly recognition. One feature that we have found about such systems is that, since the different functions and modules share a knowledge base, they allow a very integrated approach which we have termed **global control of assembly**.

Overall system

The overall system which we are investigating is depicted in Fig. 11.1 and shows the use of inference engines which have access both to a reference knowledge base and to sensor measurements of the physical world. We are particularly concerned with small batch and varied

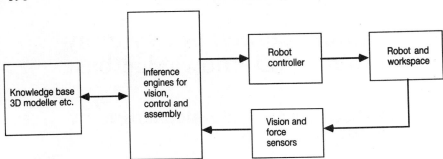

Fig. 11.1 Semi-autonomous assembly system.

assembly, where parts are initially in random locations and orientations in the workspace. We are restricting attention to parts which have been designed by a CAD system. On one hand, this is a simplification, since it means that all parts can be completely specified in the knowledge base. On the other hand, it is a practical situation since increasingly the parts in an assembly are likely to have been designed using CAD. This restriction means, in principle, that the physical world of the workspace and all events in it can be modelled in the reference world of the knowledge base. This is fundamental to the whole scheme. It is useful for research purposes by allowing the use of simulations in the development of the system. It is also of practical benefit since during assembly a comparison between sensor data and the reference model allows the detection of malfunctions, for example part inspection or loss of alignment. Thus as well as planning ideal assemblies the system will detect deviations from the ideal and instigate planning of recovery from these errors, for example the rejection of parts which fail an inspection. The project is at an early stage and in this paper we describe the requirements of each component of the system and the progress which we have made in some of them. The main components of the system are:

- Assembly planning.
- Assembly implementation covering the use of sensor information for trajectory planning and control.
- The sensor systems themselves.
- System integration.

While these are described separately they are obviously strongly interrelated.

11.2 Assembly planning

This consists of two main tasks: detailed assembly planning and coordination of assembly implementation tasks; here we concentrate on the first of these. It is being approached from two directions: developing algorithmic models for planning based on joint description, and product studies to detemine current practice. Planning based on joint description is divided into several modules.

Joint description

This includes a geometric description of matching parts from the CAD knowledge base and process information relating to the assembly process. A number of basic joints are being considered:

- Plastic press fit, as an example of a compliant joint not requiring sensor feedback.

- Peg in hole, a non-compliant joint with a single kinematic description via force feedback.

- Plane on plane, with minimum kinematic description employing vision feedback.

These are being related to a number of product studies:

- A telephone assembly with parts designed for automatic assembly.

- An aircraft equipment tray, as a fabrication example.

- A keyboard assembly as an example of multiple insertions.

Assembly description

This involves a combination of geometric and non-geometric data to describe the assembly method. The descriptions will involve motion from the geometric knowledge base, force from a process database or by self-learning, and time from the process knowledge base. This will also make use of collision-avoidance trajectory planning (see later).

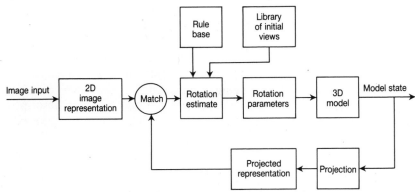

Fig. 11.2 IVISM for object recognition.

Assembly task description

This will be based upon the assembly description but will introduce the need for tooling and sensor knowledge base. It will also make use of the robot simulation model in CATIA. At present, effort is being concentrated on joint and assembly description. A knowledge base is being written in Prolog and will be incorporated in an expert system with an interface to the sensor data.

11.3 Assembly implementation

Preliminary system

In order to provide an early system for experimental purposes a method developed in Cambridge University Engineering Department for the manipulation of objects by robots for handicapped children [1,2] is being adopted as a preliminary system. This involves the use of markers placed on the objects for location by vision systems. Processing of the markers can determine the position and orientation of the objects, employing fast algorithms based on finite state machines.

IVISM—intelligent vision system integrated with a solid modeller

This method [6] is being developed to provide a means for part recognition and location by comparing views from a solid modeller in the CAD knowledge base with camera images and manipulating the solid model until the views and images match, thereby recognizing and

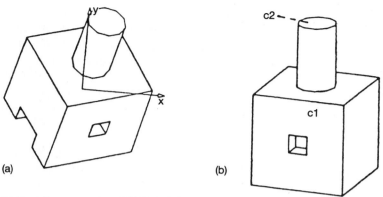

Fig. 11.3 Alignment by IVISM (after Tabandeh and Fallside [6]): (a) an initial state for tests 1 and 2, and (b) a simulated image for test 1.

locating the objects. The system is shown semi-diagrammatically in Fig. 11.2. Manipulation of the solid model is achieved by comparison of representations of the views and images using primitives with attributes. The primitives employed are closures (closed regions not totally enclosed by others) and the attributes exist at three levels:

- Attributes which are quasi-invariant with viewpoint, such as relationships among sub-closures (closures within closures): for example the number of sub-closures and their relative positions.

- Attributes which are sensitive to viewpoint, such as relative size of closures.

- More accurate shape descriptions such as adjacency of closures and approximate geometrical positions.

A frame-based system is being developed for this purpose. The comparison between the representations of views and images is carried out by a rule base which makes a rotation estimation. Initialization of the search is carried out using an auxiliary representation of the solid model which consists of a library containing a small set of projections produced by the modeller from different vantage points for each part known to it. This simplifies and reduces the run-time search. The level of detail reflects a compromise between the redundancy of a representation and the positive roles which it may play in the recognition process. At the top level the rule base is divided into modules such as structured size etc. At present the rules generally contain visibility and planar-area constraints but work is being done on other rule modules to reduce the dependency of the system on the presence of corresponded plane closures.

Fig. 11.3 is an example from IVISM. A correspondence for the most prominent closure in the image, c1, is hypothesized from the graph of the model projection. This is generally a match description and thus leads to a rotation estimation. In this case the information obtained from this most prominent closure was not sufficient and so the system focused on a second closure, c2, selected through the adjacency link in the image graph. After one iteration the object was identified and its orientation, pan and swing were identified. Work is currently in hand on interfacing the IVISM module to a vision system using edge-detection pre-processing. IVISM was developed by drawing on results from the cognitive science literature on visual perception and mental models [6]. For example IVISM uses coarse initial viewpoint changes of about 45°, following the observations of human perception described in [4]. We notice IVISM is a rule-based method for object recognition and geometric transformation. As a longer-term study we are also investigating the use of connectionist models for the same purpose [1].

Trajectory planning

Given the initial position and orientation of the part from IVISM, and the initial grasp configuration and final grasp position just prior to joint assembly from the assembly planner, the next task is to plan the trajectory in between these initial and final states. The approach being adopted is finding collision-free paths based on the 'explicit free-space' method of Lozano-Pérez and Wesley [5]. In this the moving object is shrunk to a point or, to include object rotation, the swept volume is considered, and the obstacles are grown accordingly. These grown obstacles are termed configuration obstacles or C-space obstacles. To find the shortest and safest path from initial to final state one can search a tree of paths for vertices that can 'see' each other (if and only if the obstacles and the moving object are convex). However, by representing the search space efficiently, local (nearest neighbour) goal-driven algorithms may be used to achieve the same result. This can be done by simulating wave-fronts from the starting position for the whole space and then working backwards from the destination. This method has been demonstrated so far for two-dimensional scenes. It has several advantages:

- The wave simulation can be carried out by parallel processing.
- It is not restricted to convex objects.
- The time to find a path does not increase with the number of obstacles.
- It can be extended to six degrees of freedom.

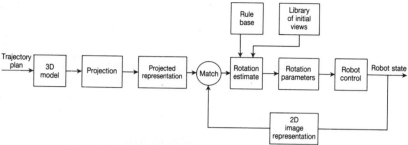

Fig. 11.4 IVISM for control by vision feedback.

This algorithm will be interfaced to a knowledge base of workspace geometry which will in its turn have its endpoint states initialized by IVISM and the assembly planner.

Control of assembly and error monitoring by vision feedback

This comprises several stages of the project. Firstly, our preliminary system will employ the marker-based vision technique which provides a basis for scene recognition and some aspects of error monitoring. Initially, this will allow trajectory planning and its implementation by feeding a sequence of set points to the existing robot controller which has conventional axes feedback. We wish to extend this to full end-of-vector control using vision feedback and employing multivariable control techniques [7]. The implementation of this is of course dependent on fast, real-time vision processing. Within the framework of the rest of the project such a control method introduces the new combination of techniques that comprise global control of assembly. This hinges on the fact that while IVISM has been developed as an intelligent knowledge-based system (IKBS) method for recognizing and locating objects controlling a solid modeller to match static views of objects in a scene, as shown in Fig. 11.2, we can also use it for closed-loop control using vision feedback. The setting for this is the provision of a control reference by the solid modeller now manipulated by the trajectory planner to provide a set of time-varying reference views. If IVISM is now used to compare these reference views with the image views obtained from the end of vector (gripper and part) the rotation output from IVISM can now be used as an error signal for closed-loop control of the robot. The significance of this is that it will deal with large misalignments of the reference and output as well as small misalignments. (Whereas a purely numerical calculation of error, say differences of area, will only apply for small misalignments.) The general form of this type of feedback control is shown in Fig. 11.4.

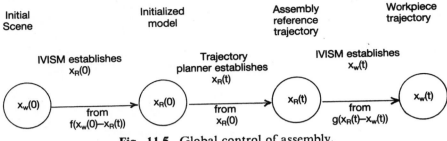

Fig. 11.5 Global control of assembly.

If we now integrate this with trajectory and assembly planning we have a form of 'global control of assembly'. Put succinctly, if $x_R(t)$ is the state of the reference three-dimensional modeller and $x_W(t)$ is the state of the scene then, this form of control is as shown in Fig. 11.5. This defines a full integration of the various components of our assembly system, as shown in Fig. 11.1 and described earlier. This appears to be a novel application of IKBS techniques in controlling assembly in a physical world from a knowledge-based reference world; it has of course close parallels with assembly by humans.

Vision systems

This falls into several parts:

- A basic framestore system with low-level image processing. This is a Seescan Devices system handling, at present, two video channels each with a $512 \times 512 \times 8$ bit framestore, accessed by a 68000-based front-end processor on VEMbus with provision for hardware processing (initially edge detection).

- A planar array of cameras for range-imaging. This is under development from a demonstration [8] using a single camera viewing a scene through a rotating Perspex block. The image is sampled at 32 equal angular increments of the block and a Hough transform is applied to establish images at different ranges. An example of this is shown in Fig. 11.6. A system based on this, employing an array of cameras and hardware processing is under design to provide real-time range-imaging for the project.

- In a separate project (*Parsifal: Alvey project 074*), an Alvey consortium is developing a 64-Transputer parallel processing system. This will be employed in the present project to provide fast parallel-processing for several of the component studies, including vision.

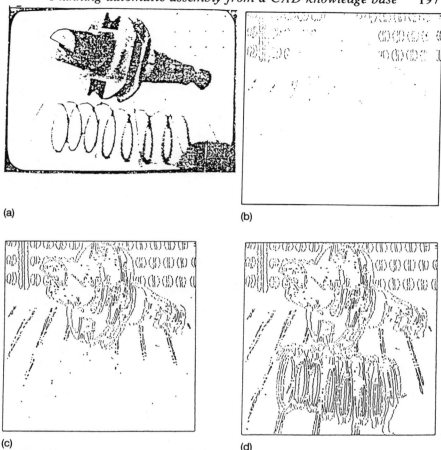

Fig. 11.6 Range imaging results (after Wright [8]): (a) a sample test image, and (b) the range image thresholded in depth showing the acoustic tile backdrop to the test scene. In (c), the image of (b) is shown thresholded at a close range to add the machine tool cutter, and in (d) it is thresholded at minimum range to add the spring.

11.4 System integration

This will take place in several stages: first, software integration of the separate modules. This will initially be carried out with non-real-time performance. As faster processing becomes available the integrated system will move nearer to real-time performance and allow a fuller study of aspects such as end-of-vector control and error recovery. The aim is to develop the principles of the system in an integrated context.

11.5 Conclusions

This paper has discussed plans for the framework of an automatic robotic assembly system linked to a knowledge base and given early results for some of its components. It is believed that it indicates one feasible solution for a particular type of system characterized by the assembly of small-batch, varied assembly tasks from random initial dispositions of parts. It has discussed the aims and function of the various components of the system and some aspects of their integration. A particular integrated approach termed global control of assembly has been put forward which is believed may be applicable in other types of semi-autonomous knowledge-based systems employing vision sensing.

Acknowledgements

This work forms part of SERC ACME project GR/D 71016. It is also supported by our industrial collaborators, IBM UK Laboratories, Seescan Devices, and British Aerospace, Preston whose assistance we gratefully acknowledge. We would also like to acknowledge the part played by other members of the project team in making this paper possible.

References

1. F. Fallside and L.-W. Chan, 'Connectionist models and geometric reasoning', in *Geometric Reasoning* (Proceedings of a conference held at the IBM UK Scientific Centre, Winchester, December 1986), OUP, 1989.

2. A. Ginige, 'A vision system for a robot working in a semi-structured environment', PhD Dissertation, Cambridge University, 1986.

3. A. Ginige, W.S. Harwin and R.D. Jackson, 'Finite state machine to detect markers', Proceedings of an International Conference on Image Processing, IEE, London, 1986.

4. M.A. Just and P.A. Carpenter, 'Cognitive coordinate systems: accounts of mental rotation and individual differences in spatial ability', *Psychological Review* **92**,2 (137-172), April 1985.

5. T. Lozano-Pérez and M.A. Wesley, 'An algorithm for planning collision-free paths among polyhedral obstacles', *Communications of the ACM* **22**,10 (560-570), 1979.

6. A.S. Tabandeh and F. Fallside, 'AI techniques and concepts for the integration of robot vision and 3D solid modellers', Proceedings of an International Conference on Intelligent Autonomous Systems, Amsterdam, December 1986.

7.   S.J. Williams, 'Frequency response multivariable design for robots', Proceedings of an IEE Colloquium on Robotic Control, London 1983.

8.   S. Wright, 'Hough transform analysis of data from a planar array image sensor', Proceedings of an Alvey Conference on Computer Vision, Bristol, 1986.

Discussion

(This paper was presented by Fallside)

Stobart: This planning-for-vision issue is quite an interesting one. I think that it was probably first explored by Lozano-Pérez and others who implemented a solid modeller called GDP at the IBM Thomas J. Watson Research Laboratories [10]. The aim was to predict the likely configurations of an object by considering where the centre of gravity was and identifying planar faces on which the object could rest. And what it did was to break the configurations down to the most likely ones, in descending order. It has been further pursued by Baumann at McDonnell-Douglas [9]. Was your technique similar? If it was different, why was it different?

Fallside: It was different for a number of reasons. We are concerned with object recognition for objects which have been produced by CAD. So we were being forced in two directions, such as the aircraft part or the computer keyboard. The primitives that we chose were influenced by these directions; we knew that we were going to meet objects like those. I think that the rules which are being applied, and the use of frames, are different from their work.

Stobart: GDP is a faceted modeller, so that non-planar faces have to be represented approximately. The IBM team's problem was made much easier by considering configurations where one of the plane faces was in contact with the assembly surface. They could compute where the centre of gravity was, and hence the stable configurations. I think that if you had a solid modeller which had quadric or other curved surfaces, you'd have to approximate it.

I have another question on assembly description techniques; this is an area that is of interest to me. I think that the solid modeller is perhaps going to change the way in which a designer designs an assembly. In particular, in engine design, the approach is build the general assembly, and then to detail the parts; whereas what we might do with a solid modeller is perhaps to conceive of an engine with its rough dimensions, and then build up the parts and finally assemble them. The assembly description is not simply a description of how they will fit together; it carries certain technological information: fits, clearances, surface finishes and so on.

Fallside: We see this as something that is starting off broadly as geometrical descriptions, plus a number of cases from past experience of the actual joining activity; in effect, we are trying to pick out a number of canonical joints. We expect that we will have to expand on these in time, but we hope that the system will be able to learn, by looking at the family of joints, and attempting to break it down into canonical joints, if there is a limited set of them; that may have wandered a little from your question. I think your other point is quite true; we are interested in—clearly our industrial partners are interested in—the impact of this sort of assembly on design. If you are designing for a particular purpose, then this includes its assembly, and one may have some effect on the other.

Orr: How do you plan to represent different views in your library of views of the object? What criteria will you use for getting the best matches?

Fallside: We view this as a dynamic activity. I mentioned a library of initial views; these we will certainly store, like the directory to the models themselves: to locate likely models quickly. But once you have homed in on the appropriate model, say the one for a particular tool, you're not then using stored data, you are using data which is generated from the modeller, which has previously been set up during the computer-aided design of that tool. We view that as stored information which is available in that way to the model, which we then use at the time.

Orr: What form will these stored views take?

Fallside: Well, this goes back to a point that was made earlier about the representation. Do you in fact go through a procedure which will label your representation, or do you include that labelling when you are doing design? There is no reason why you shouldn't do that. Your model information could have a fair amount of hand labelling in it, which would say where the major primitives are. I think that the treatment of the labelling problem for the image data is slightly different because, clearly, it's noisy.

Williams: Do you have any feel for when you will change from the symbolic data representing the part to the explicit pattern data? Have you got any feeling for how close you are to a match when you make that transition?

Fallside: It something about small displacements. If you consider that you're near a match, it's far enough away so you still have a stable equilibrium when you are matched. Clearly, with non-trivial shapes, as soon as you get far away you are going to come into other minima, so the whole system might be unstable, because it was seeking a minimum which was not an alignment in purely numerical terms. The coarse rotations that we're using at present are about 45°, these are the changes that we're using at level 1, the viewpoint invariant ones. This seems to be appropriate. That's something about the symbolic end. The change from that to pattern or numerical matching—it's difficult to tell—somewhere less that 10°.

Williams: How many pixels is 10°?

Fallside: That's not really the point: it's shape-dependent. You could dream up some objects where the stability of the match would occur for a very small change, and then other objects where you would remain within that stable region for a very large variation in alignment. Certainly, it's around 10°.

Jared: If I may ask a slightly convoluted question. You said that you are starting off on this project in the hope of better technology arriving in order to approach real-time performance. You are making quite stringent demands on your solid modeller in the view-matching process, and you also mentioned perhaps improving your solid modeller using parallel processing. This has obviously got a lot of implications for the representations in the modeller, whether they're the ones that I talked about, or others yet to be invented. Are you confident that you are staring on a path which is not going to lead to insoluble problems? Might it not be better to consider the question of modeller representations in somewhat more detail—what they might be able to do for you, and what is clearly impossible—now, rather than finding things provably impossible when you've got quite a little way into the project? You might end up taking a slightly different direction.

Fallside: That is a very good point. I stressed the related problem in respect of hardware for vision processing; but you're absolutely right that the technology of modellers with this capability is limited. You know much better than I do that this is stretching existing solid modellers. We are really approaching this problem in two ways. Again, this comes back the fact that we are not producing a fast gizmo that, in three years time, will do real-time semi-autonomous assembly of one object. We're interested in the general problems, and one of the general problems is; what is the form of the modeller that will do that? We're approaching this really at two levels.

One is the 'fail-safe' method, which involves putting a marker on the object. This will produce, using vision, a special case version of the system that will recognize objects and their orientation. It will allow the testing and planning of assembly, and closed-loop visual feedback as well if we want it to. So we will not find that the whole project comes to a halt because we meet a major problem in the modeller; we will carry on, using this other method.

In the meantime, if we discovered that we were overstretching available modellers, we would certainly spend time on a special case study for a modeller which wasn't general purpose, but had the capability of looking, as a sort of experimental tool, at this form of recognition and of control, and I would stimulate as many solid modelling people around to assist on this, and to generalize it. It seems to me there are some general problems to be attacked in this.

Sabin: I think that you're being quite unnecessarily defensive on this point. One has to distinguish between software which builds the model in the first place, which is where the cleverness is, and the modeller which is just giving you different views of it. You could use—for instance—a Silicon Graphics display simply to turn something around in almost real time. That capability certainly exists, but the model there is just a collection of facets. From your really clever model-building software, all you have to do is to produce a set of facets, and from there on special-purpose silicon—we know—can generate views which can be compared with what the camera sees. I think that you might not be on quite such weak ground as you think.

Jared: He really doesn't need a solid modeller to do what he's doing, as he said himself; using markers, a solid modeller is not necessary.

Sabin: He needs a collection of facets, but a solid modeller is quite a convenient way of building them.

Fallside: I was certainly being cautious; there are many different views on these matters. I think that there are two points. We would certainly wish to use a CAD system which is widely used by industry for this sort of purpose. Then, if you produce an assembly system of this kind, you know both that you will be dealing with objects that have been designed by the system, and that it is something which is known in industry and would be usable. But one has got to back off from that—and we have already done so—in going towards simplified modellers. We will certainly not abandon the use of solid modellers because we can't get a general-purpose one, which will do all the modelling that we need to do.

Additional references

9.   E.W. Baumann, 'CAD model input for robotic sensory systems', Proceedings of AUTOFACT 4, Philadelphia, 1982.

10.   M.A. Wesley, T. Lozano-Pérez, L.I. Lieberman, M.A. Lavin and D.D. Grossman, 'A geometric modelling system for automated mechanical assembly', *IBM Journal of Research and Development* **24**, 1 (64-74), 1980.

12 Part representation in process planning for complex components

P. HUSBANDS, F. MILL and S. WARRINGTON

12.1 Introduction

This paper deals with the problems faced in describing components to process-planning facilities in the context of discrete part manufacture.

In manual process planning of mechanical components the planner is usually given a drawing of the part to be made, and then uses this two-dimensional line representation as the basic input to the planning process. The planner reasons about the part's shape and then develops a three-dimensional or partial representation in his head. The planner does this by identifying certain problem areas or by seeking familiar patterns, such as slots or threaded holes, which he can readily identify with machining methods. These patterns may be called **features** and the parameters which dictate size or tolerance may be termed **attributes**. Problems can be considered by looking at the relationships between individual features, for example due to geometrical tolerances or the fact that one feature might impede tool access to another.

The original designer's conception of a part is quite likely to be in terms of three-dimensional features such as slots, holes or splined shafts. It seems unfortunate that the design/planning process starts with a meaningful three-dimensional feature-based representation which is then translated to a two-dimensional line drawing by the designer and which is again translated into a different three-dimensional feature representation by the process planner. However, the features used by the designer and the process planner do appear to differ in their nature. Designers seem to use functional features whereas the process planner uses manufacturing features. An example of this might be a designer requiring a groove on a component to allow the flow of some fluid. It is the cross-sectional area of the groove that is important, to ensure that a predetermined flow rate is maintained. The process planner, on the other hand, is probably going to be most interested in the shape and location of the groove and may well differentiate round-bottomed from flat-bottomed grooves as separate types of feature altogether.

As well as having the drawing itself, a process planner is likely to have other information as an input to the planning process. Such information may consist of material type, batch size or blank shape; or the planner may generate input by applying a classification and coding system which would yield a part code. This enables the planner to find other components in a database which have similar codes and therefore (hopefully) similar manufacturing plans. This approach is used in variant-type computer-aided process planning (CAPP) systems. Such systems take codes (e.g. Brisch or Opitz) as their input and automatically retrieve plans for similar parts. These plans may then be edited on the screen and a new plan output with relatively little effort. They are, however, limited in their ability to plan for certain firms' needs and for certain types of components and some systems use codes with as few as seven or eight alphanumeric digits. If process planning is to be carried out automatically and without reference to another component which happens to have the same shape, then a more complete description must be used.

12.2 Requirements of a perfect representation for process planning

There are many requirements of a perfect component representation for process-planning systems. First, the representations produced must be complete in process-planning terms; in other words they must contain information on the relationships between various parts of a component such as those specified by geometrical tolerances. The representation need not contain the complete geometry of the part but might instead, for example, use parameterized information about the nature of a surface. A complex surface might then be labelled as 'complex' and various attributes noted, such as the minimum curvature on the surface or simply the fact that the surface is not convex. Alternatively the surface might be classified according to some pre-defined system. The second requirement of a representation is that it should be flexible. It is likely that future development of any CAPP system will always demand that new features should be put into the system. This might be due to installation in a new site or to new processes becoming available which enable the use of new features. Alternatively, changes to the design procedures used in a particular plant are likely to demand new features from a purely functional viewpoint. Thus the general representation method must be flexible to allow for updates and extensions as the user may wish. It is also very important that a representation should be easy to edit. If the model is being used to

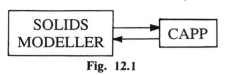

Fig. 12.1

describe intermediate states between component and blank material then the CAPP system must be able to delete features and add new ones as desired as well as updating changes to existing features. In practice it is also important that the system developer or administrator should be able to read representations and possibly interact with them. It may then be possible to create models of seemingly impossible components: for example, a single face. This could save the modelling of a very complex casting if, for instance, only a single flange needs to be machined.

One of the most important points about a representation which is often overlooked is that it should be capable of fully describing the interrelationships which exist between the features of a component. A component cannot simply be regarded as a collection of individual independent features. This would imply that a process-planning system could be applied to single features on their own and then a list of all the single operations could make up a complete process plan. In fact most of the problems in process planning occur as a result of interrelationships in the component. For example, overhangs can prevent tool access to the desired features, or a seemingly simple feature may be tied to several other features via a complex geometrical tolerancing scheme [1]. This often happens when holes have to be jig-bored instead of simply drilled because of a required positional tolerance. Any serious representation technique must be capable of fully describing these sorts of tolerances as well as those describing factors which affect machining anteriorities.

Lastly it is important that the representation which is to be used allows the operation of powerful production rules. If a process-planning planning system is to be completely based on knowledge-based systems (KBS) techniques then this consideration is crucial.

If powerful CAPP systems can be developed using KBS methods then it is likely that any part-representation technique would necessitate the generation of large models for relatively simple components. This would have the effect of moving work upstream in the design/planning process. The production engineer would no longer have to spend time developing process plans but would instead have to spend the same (or probably more) time developing the initial representation of the part which would then have to be input to the CAPP system. It is therefore important in the long term that a representation should be capable of being generated automatically (at present in theory at least) from some CAD representation of the part.

12.3 Automated generation of part representations

At present there appear to be three approaches which can be taken to solve the problem of automatically creating representations. The first method would involve using a solid modeller directly as shown in Fig. 12.1. In this scheme information is passed between the solid model and the CAPP; that is, the solid model is interrogated and updated by the CAPP system as planning proceeds. This approach is indeed complex and lacks many of the desired attributes discussed in the previous section (such as geometrical tolerance information). It may however be possible to use such an approach in areas where the component type is well defined. Such systems do not yet exist, to the authors' knowledge, but are being proposed by several researchers in process planning.

The second approach, pioneered by Kyprianou [4] and later discussed by Jared [3] and Henderson [2] would create a secondary representation which might then act as an interface to a process-planning system. Others have worked in this area and to date these methods have been built on feature-based representations. A schematic diagram of this approach is shown in Fig. 12.2 and is discussed more fully in the papers cited.

The third approach does not involve the extraction of features from solid modellers but instead concentrates on creating full feature descriptions at the same time as the creation of a solid model. This is the approach which is currently being investigated by the author. Its main attribute is that it has a separate interface to the user as shown in Fig. 12.3. This means that the feature description is not related to the solid model and that the interface may be used to generate other models, for example simpler three-dimensional wire frames. The use of solid modellers for everyday design work in the engineering industry is by no means widespread and cannot be guaranteed to be so in the near future and so there are reasons why linking research work in process planning to solid modelling is not necessarily beneficial. Furthermore, it is possible to argue that if solid modelling were to become more acceptable to designers it is likely that it would be used for single components which had special problems (i.e. in the same way as finite element analysis). These components would not be likely to form a large proportion of the parts to be planned.

There are, however, many problems in trying to use this approach, the most important of which is that giving designers set of features to work from does limit the scope of things which they can actually design. This is currently a controversial situation in which designers often argue that such systems would place unacceptable constraints on them while production engineers argue that constraining designers to make

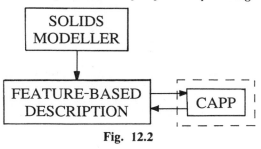

Fig. 12.2

components which are easy to manufacture is reasonable in cost-conscious environments. Another major problem with this current technique lies in the implementation of such a system. The need to be able to undo mistakes which might be made is important but would be difficult to achieve in practice.

A demonstration of this front end has been developed and is to be rewritten so that an improved man-machine interface can be included along with some form of error recovery. The present simple system drives the BOXER solid modeller and creates the information in the required semantic and syntactic form for driving the knowledge-based process-planning system.

12.4 Part representation

Development of a suitable representation system has taken several years; in this section that work and a description of the present system are outlined.

Early attempts at using feature-based parametric descriptions were too simplistic in many ways to be able to handle realistic components. Initially simple lists were used to give the attributes and parameters of a particular feature [6]. This method, however, could not handle complex relationships between the features of a component and so the lists were arranged in a hierarchy with the top-level feature being a **component**. This had attributes such as material and batch size, for example. Child features of the component were, for instance, slots, grooves, pockets or simple surfaces. The problem with this approach is that it is inflexible and very often there is a desire to have a child feature which itself wants to be a parent of some higher-level feature. Finally a network approach was developed whereby the relationships between features are declared explicitly. The representation now being used consists of a binary relational database where each entry consists of a triple of the form:

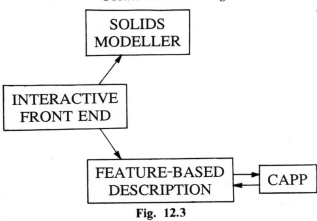

Fig. 12.3

<entity> <relation> <entity> .

This differs from standard networks in that there is no real distinction between entities and attributes. Some 40 + relations have so far been defined with around the same number of entities. these entities may be features, names, or variables. Perhaps the most basic relationship is described by the 'is_a' statement which allows one to say what type of feature is being described, for example:

A is_a plane.
H is_a thro_slot.

The use of binary relations has advantages over other database methods in an application such as this partly because a consistent framework can be used throughout the total planning system. Binary relations are used to describe the component, the blank and the machine and tool capabilities in a consistent way and this helps in the development of application software. Rules in the system may use binary relations as antecedents and consequents and the programmer accessing the database needs to know little about the file design. He does not concern himself with records and the programmer does not have to choose keys or relegate data to fields. This is particularly important in a development system where the final requirements cannot be completely foreseen.

The representation technique allows the use of some higher-level operators or relations which treat a triple as an entity in itself and hence allow some longer relationships. An example of this is where a qualification of an initial relation is required. Consider, for example, the part of a component shown in Fig. 12.4, where two faces A and B are related by a geometrical tolerance. This would be represented by the statement:

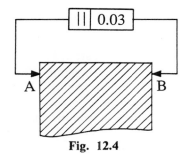

Fig. 12.4

A para B with_tol 0.03.

There are some problems in using this approach, for example multi-attribute retrieval might prove inefficient. In the present system it is envisaged that such searches should be rare. An example of such a search might be 'to find all planes which have surface tolerances, parallel tolerances to holes, and which are hidden by other features due to overhangs'. There is currently much interest in the use of binary relations in database technology within the computer science community and some workers have produced specialized hardware for fast storage and retrieval of triples, for example the 'intelligent file store' of Lavington et al. [5]. One way to improve access times generally would be to use dynamic hash tables but this involves duplication of data and in the present form the representation would already use up some tens of kilobytes for relatively simple components.

12.5 Conclusions

The use of the current system of binary relations which express the features of a component as well as the relationships between them is at present proving successful. Although many parts can be modelled with the current system it is very time-consuming to model a part manually and there is currently a great need for a system which can generate part descriptions automatically if a knowledge-based systems approach to process planning is to succeed.

Acknowledgements

The authors would like to thank Forbes Simpson for his work in developing the prototype feature-based solid modelling driver. The work is funded under SERC grant GR/D 63101.

References

1. *Geometrical Tolerancing*, BS 308 Part 3, 1972.
2. M. Henderson, 'Features recognition in geometric modeling', Proceedings of the CAM-I 13th Annual Meeting, Arlington, Texas (5:1-5:13), 1984.
3. G.E.M. Jared, 'Shape features in geometric modeling', in *Solid Modeling by Computers* (M.S. Pickett and J.W. Boyse, eds), Plenum, New York (121-133), 1984.
4. L.K. Kyprianou, 'Shape classification in computer-aided design', PhD Dissertation, University of Cambridge, 1980.
5. S. Lavington, M. Standring, and G.B. Rubner, 'A 4 Mbyte associative predicate store', Proceedings of the 1st Alvey Workshop on Architectures for Large Knowledge Bases, Manchester, 1984.
6. F. Mill, and S. Spraggett, 'Artificial intelligence for production planning', *Computer-Aided Engineering Journal* 1,7 (210-213), 1984.

Discussion

Stobart: One thing about your representations; it is possible to have two different items of data meaning the same thing. This implies a degree of redundancy in your database. Have you taken steps to minimize the problems of redundancy?

Mill: No. The approach that we took to the problem of redundancy was to say 'Who cares really?'. The other needs or requirements from the system were such that we thought they were far more important that the redundancy. There are various stages in the representation where there is redundancy. (We don't have a very consistent approach in the representation either, I might add.) Here's an example of one of the most obvious ones. If we wish to relate two pieces of data A and B, we may have to include a relation of the form (B,A) as well as (A,B); that's a ridiculous level of redundancy. We don't bother about that; we just take account of it in our searching procedure; we look for it in both positions. There are areas, where we are dealing with edges and so on, where we just state it twice, and say, well, it doesn't really matter very much.

If at some stage we get into problems of redundancy, where we've got too much data and it's starting to become a bit of a clutter, I think then we might try to reduce the size of the representation and to write some rules to take account of it. At this stage we didn't really see a need to do it, that's all. I might add that we're prototyping, so when we're trying to rethink it all again, we'll probably discuss those problems further.

Todd: Are you putting in the redundancy yourself, or are you getting rules to generate the redundancy anyway?

Mill: What we're doing is to try to write some simple process-planning software and, while we're doing it, if the software says that it would like to know if there is a concave edge belonging to this plane, then what we do is to go away to the representation and to write in whether there is or not. That might sound as if it's cheating a wee bit. It is if you were just planning for that single component, but the idea is that you get a list of the things that you explicitly stated. What we might do in the future is to try to get some sort of minimum representation, and then apply some sort of rule base to infer other pieces of information, which we can then include into the representation.

Sabin: The problem with redundancy is usually that it leads to potential inconsistency. From what you're saying, I suspect that, if you did get an inconsistent database, it wouldn't be the inconsistency that would be the problem, you'd just get wrong answers, whichever end was wrong.

Mill: Yes, that's right.

Sabin: But could you give us an example or two of the kind of processes that 'dig at' the representation, and of the kind of detail that the representation contains.

Mill: What we initially tried to do with things like processes is this: we don't look at creating a slot as a single process, although a hole might be treated in this way. But the idea is that you get some sort of strategy for creating slots, and within that strategy you may have to look at various processes. You might use more than one process for a feature.

We don't bother too much about the processes themselves, we try and look straight at the machines; there are two levels there. What we try to do is to describe the machines in exactly the same form as the database. If you had a hole to drill, for example, with a set of tolerances, you would look for a machine that was able to drill a hole with that tolerance. It is a matching process—using an estimate first. With a drilling machine, you would have a list of features that it could create. Similarly, with things like milling you would have a list of slots, pockets and so on. It would be just like the part description, but it would belong to each machine.

Sabin: So you might discover at that stage that you could either mill that slot or create it using EDM (electro-discharge machining).

Mill: Yes.

Sabin: What kind of logic would you be applying to decide which?

Mill: That really is something that we're constantly discussing, and really haven't done very much of. The basis behind it is to say—the first pass, if you like—the first thing we want out of this is a list of possibilities—so you might develop something saying that you can EDM it, you can mill it, or various other things. Later on in the program, we make the decision whether it should be EDM machined or milled.

The decision is based in a set of weighting factors which we've called 'costs'; but they're not straight costs—you couldn't use them for estimating.

For a hole, for example, it's preferable to drill it rather than produce it by EDM, unless you have to, because EDM is much more expensive. The other thing is that, we're just using arbitrary numbers here, but you try to pick out say three possibilities (it could be thirty-three, we don't really know the best number yet), and then later on, your software has some sort of optimizing problem to determine which one of the three it should be.

It's based on, firstly, the cost for making that feature on that machine, and secondly upon the relationship between that and other features. So, for example, if at some stage I've got a simple hole here and a machining centre or a hand electric drill that can make the hole. Obviously the electric drill is much cheaper, but if everything else that is being done to the component is being done on the machining centre, then you don't want to go to all the bother of taking it off and drilling it, and then maybe having to put it back on to get other operations done to it. The costs are the simple costs and the movement costs, which must be added in.

Bowyer: This isn't really a question, more a consolation. If you do have an object with internal inconsistencies, presumably your difference procedure could be used to operate on it and a copy of itself in order to resolve, or at least report on, inconsistencies?

Mill: Yes. It couldn't resolve them, but it could report them to you. A lot of the inconsistencies that we've had so far just stem from the fact that it's a human being who is trying to write down these representations, who is forever labelling surfaces wrong and so on.

Bowyer: Do you think that there is any potential to automate that, from a solid model?

Mill: It's something that we're actually trying just now to do using the Leeds University NONAME modeller. We have only performed some very very simple experiments on it. Suppose we've got a block and want to put a slot in it. The program asks a series of questions, and the designer would then say that he wants a slot in it. (I say designer, but it is really an input to the process planning and not a design solid modeller.) The system then goes away and creates another block and subtracts that block from the initial block. It's not all that difficult to do that, but you have some other information to type in at the beginning, to label faces, and that sort of thing. We've looked at it from that point of view. As Jared said [7], you end up with a whole list of features which the designer can use. If you are working, as I am, in a department with mechanical engineers, they throw their hands up in horror if you restrict a designer to using slots and holes with threads in, and things like that. They all think at a much higher level than that—so they say; I'm not fully convinced about it. I think that a lot of the design work that goes on in industry is probably down at a feature level. If that is the case, then, yes, it might be possible to automate that kind of thing. We've only done some experiments on it so far.

Wright: Designers are supposed to think about manufacturability when they're designing a part. Some of the hard problems that you have described about taking decisions: you're really talking about trying to infer what was in a designer's mind in the first place. Isn't there some case for having a more interactive system at the design stage, where you're actually interrogating him more about what would be sensible from a manufacturing point of view?

Mill: I don't thing we are trying to infer what is in the designer's mind. I can see that there are areas where it comes close to doing that, but surely what you are trying to do is to infer what should be in the process planner's mind, if you like?

Wright: I was thinking of things like the choice of machining operation. The designer must think, when he creates a feature, that he is making a stringent requirement that will need an expensive process.

Mill: Some of them do, but some designers claim that they don't. Some designers claim that you shouldn't think about machining at the design stage. I actually heard someone say that in a discussion we were having a couple of weeks ago. He is a professor in the subject, and views design as a very high-level sort of thing, which shouldn't be interrupted by thinking about manufacturing constraints.

I should add that one of the problems that we've had is that going back to try and get information from a solid modeller is likely to be very difficult if it's on a conversational basis. So what we have been trying to do is to create a list of features while you're creating the solid model; that isn't quite the same. It also means that it's a one-way process: once a designer has done a design, we're accepting that it's perfect, in terms of functionality. So we're going to go away and make it. One of the things that I have thought about, is that, with all this knowledge and work that has gone into the various process-planning activities, there's no reason why we shouldn't take that out and give it to the designer anyway, and at least he can use it, and you should get better designs, but there won't be any tie-up between the two.

I might just at this stage refer to another project that is going on in my department. Some people looking purely at the problems of process selection. They are developing a small expert system to which you can describe your symptoms—feature and its attributes—and it suggests a few processes. I think that is the sort of thing that a designer would be able to use. I think that we ought to be able to develop something similar, based on the knowledge that we've put into this work.

Sabin: Just a point: people are only people. A designer probably has a difficult enough task designing something that works. A colleague at Birmingham University knows of more than 250 different material removal processes. Asking for designers to be up-to-date in all the capabilities of modern manufacturing technology is just not realistic.

Mill: I think the thing is that you can't trust the designer. The designer may have had it in mind to use some particular process for a a particular feature, but you may think that you have a newer or better technology than the one that which the designer was thinking about. You don't want to think too much about what the designer wants, in that respect.

Wright: I suppose that you could link process selection software interactively with your program and alert the designer to the cost of the component which he's just specified; and ask him whether he really wanted to have that very expensive feature or not; and would he like to think again?

Mill: My view on that is that it all depends on how the technology turns out in the end. If, in five years time, we have a nice process-planning system which is working and process-plans things almost instantly, then it would be very cheap to put that that process-planning system up on the designer's desk, as well, and he could just look at it as a matter of 'what if?'. But I really don't know about that.

Sabin: To return to this business of the features being the same or similar: the actual function which the designer's designing for is almost always related to the material that is left after machining—the space between the things that you machine out. Now, there are things like slots which you do think of in that way at design time, but the slot, from the point of view of the designer, is a mechanism to hold something between two sides. There will be small differences which keep cropping up, as well as the big differences between things like webs and pockets.

Mill: There are things like an oil wick. A designer may see a little groove with a flat or a curved bottom; he's not really interested in things like that; he thinks about things in a slightly different way.

Wright: One of the reasons that I am interested in this is because we recently did a study on design for manufacturability of some components in assemblies that we'd been making, and we discovered that, on an assembly which was about 500mm long by 300mm wide, the designer had managed to specify about 120 different sorts of fastener. Although it's hard to get the designer to think about manufacturability, we have to start getting better at it. Maybe the work that you're doing is one step in that direction.

Mill: Yes, I completely agree with you. We have to start trying to push designers in that direction. Just how we do it exactly I'm not too sure.

Irving: I wonder whether you used value analysis at all as one way out of this dilemma?

Mill: My own opinion about value analysis is that it is a technique that designers should use anyway. It's not really something that we're going to look at in any of this work. In my discussions with design engineers—I'm not one, I'm a production engineer, not a mechanical engineer—I find that there's a lot of controversy about this question. Some of them are very interested in designing for manufacture. Other eminent design engineers have never heard of value analysis; that's shocking, but it's true. I think that the only way that we can start to do that is on an educational basis. Start teaching our students about it, push it in the press, and hope that designers will take up value analysis.

Additional reference

7.	G.E.M. Jared, 'Recognizing and using geometric features' in *Geometric Reasoning* (Proceedings of a conference held at the IBM UK Scientific Centre, Winchester, December 1986), OUP, 1989.

13 Knowledge-based systems in process planning and assembly design

J. L. MURRAY and M. H. MILLER

13.1 Introduction

Engineers are concerned with the creation and manufacture of products. In general a product consists of a number of components and sub-assemblies linked in a unique manner to satisfy a particular consumer need; each component can be described unambiguously in terms of its geometry and material. Currently there is considerable interest in the development of knowledge-based, or expert, systems to assist in both the design and manufacturing phases of products.

Although some of the simpler decision-making processes, such as material selection, may be handled by knowledge-based systems with no geometric capability, much of the decision making requires some knowledge of, and reasoning about, the geometry of the artefacts to be produced. This affects the design phase in such areas as component selection and tolerance determination and the manufacturing phase in aspects including process planning and robotic assembly.

The obvious way of handling geometric information is through the use of three-dimensional modelling systems. There are three fundamentally different approaches used by modellers, wire-frame, surface and solid; of these solid modelling is the only one which provides an unambiguous representation of the geometry.

Unfortunately, the capabilities of existing modelling systems are limited from the point of view of linkage to and use by a knowledge-based system. For example, suppose that one wishes to construct a system which needs to interrogate and manipulate a solid model via a three-dimensional modelling system. The potential for such interaction provided by current modelling systems is very limited.

Clearly if knowledge-based systems are to be developed for effective decision-making in CADCAM, it is desirable that three-dimensional modelling systems should change. In particular, it is important that more of the lower-level geometric knowledge is moved into the modeller so that it can reason about aspects of the geometry of the model, such as the classification of features and the storage and manipulation of the relationships between components of an assembly. It is also essential to provide a more adaptable interface by which knowledge-based systems can communicate with such modellers.

This paper looks at two applications with requirements for more intelligent three-dimensional modelling systems. The first is in the area of process planning and the second in the design of assemblies.

13.2 A knowledge-based system for process planning

Background

Operations planning is the task of selecting and sequencing machines and processes whereas **process planning** involves the detailed technological selection of individual operations on each machine. The two together are fairly complex tasks which until now have been performed exclusively by human planners. However, for several reasons there is growing interest in creating knowledge-based systems to assist the planner in these tasks. For example:

- Van't Erve and Kals [5] report that, as a result of decreasing product lifetimes and increasing variety of products, batch sizes in part manufacturing have been falling steadily. This places increasing pressure on planners in order to maintain the level of production.

- Davies et al. [3] note that in many companies the average age of process planners is over fifty and that soon there will be a shortage of process planners with the necessary knowledge.

- The knowledge used by process planners represents years of accumulated experience. As planners retire their expertise will gradually disappear. The capture and formalization of such knowledge could stop this erosion and enable the knowledge to be extended.

In view of this and the fact that these processes can be costly and prone to error the authors are collaborating with Ferranti Infographics and Ferranti Defence Systems in the development of knowledge-based systems which will assist planners in these processes. The first phase of this venture involved the development and evaluation of a fully automatic machining system for the data generation and manufacture of precision 2½D milled parts [1]. The current programme of work is aimed at extending this system through a knowledge-based approach which involves linking a knowledge-based system with a solid modeller.

Design discipline

The planning and programming systems outlined in the following sections must been seen against the design and manufacturing environment of Ferranti Defence Systems.

Within the company a strict code of design discipline is adhered to. Standardization is regarded as essential and plays a key role in facilitating the design of components capable of performing complex functions. Wherever possible, the parts are constrained to be manufactured from cuboidal blanks of pre-stretched aluminium alloy machined from one or two opposing faces using cutters limited to four diameters and three lengths. Non-standard machining operations are strongly discouraged although in some cases they are essential. Complex structures or components may be built by riveting or dip-brazing an assembly of 2½D parts. In manufacture the range of machine tools, cutters and clamping systems are all minimized.

Many of the components manufactured are complex in shape, produced to tight tolerances and made in small batches. A cellular or group technology approach has been effectively implemented. During the development stage of a product, design changes are frequent and the lead time short. This places exacting demands on the planners who need to respond quickly and accurately to changes in design.

13.3 The automatic machining program (AMP)

Part-programmers generally set about their task by examining the detailed drawing, deciding on the method of holding and the machining sequence, and then extracting the dimensional information required to determine the cutter path. Computer programs which assist part-programmers will generally automate the geometry extraction phase, leaving the other decisions to be taken by the programmer on the basis of experience or calculations. The resulting solution is very much dependent on the part-programmer's ability from the point of view of both accuracy and efficiency.

With the strict enforcement of standard approaches, the decision regarding the methods of holding is simplified as the work-holding strategy reduces to a choice between elementary clamping and bolting to a standard platen. Some ten years ago it was decided to try to capitalize further on the improved design discipline to develop a package which automates the planning and part-programming processes for 2½D parts. At the time it was perceived that, although there had been a number of attempts (e.g. the CUBIC system developed by Stolenkamp et al. [4] and the COBASH system of Bosch [2]), no suitable commercial system existed for dealing with prismatic parts. The intention was to develop a fully automated planning and programming package which would generate complete manufacturing information from an input file including part geometry, material and inspection data. The package would derive:

- The group technology cell most appropriate for the manufacture of the part.

- The blank size selection from a standard range.

- The sequence of manufacturing and inspection operations required together with the estimated completion times for each operation.

- The best method of component clamping of the blank and its location on the platen.

- A full package of information for the validation of the selected process and for issue to the shop-floor for part manufacture and inspection.

Although it was hoped that the geometry of the component could be captured from design layouts, in general a detailed drawing provided a more suitable starting point. the geometry is expressed as a series of closed or open 2D profiles, starting with the outer boundary of the part. With each profile is associated a pair of depths (Z coordinates) which determine its vertical extent. In addition the position of each hole or group of holes is specified.

Tolerance data and notes can be appended. In order to verify that the input data are correct, check plots are generated for each face of the component. An example of typical output produced by the system is given in Fig. 13.1. If requested by the user, inspection features can be highlighted on the plots.

13.4 Machining strategy used in AMP

A modular approach has been followed in developing the machining strategy for components. This can be regarded as being made up of six modules.

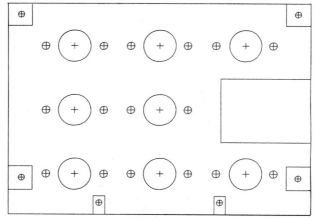

Fig. 13.1

Method of holding

Three methods of blank and workpiece holding were investigated: machine clamps, vice and bolting. Of these, bolting to standard platens incorporating a grid of tapped and reamed holes was generally preferred with machine clamps as a permitted alternative. Although the former approach necessitates larger blanks, it removes the problem of clamp relocation and avoidance.

Blank sizing

When blank dimensions are computed, allowances are made for the method of holding selected, together with the most appropriate cutter diameter for parting-off the finished component.

Profile division

The initial task of this module is to determine the location of the attachment points of the finished machined component to the blank frame. In the case of a complex part, different strategies may apply giving rise to different solutions to the problem. The choice is then made on the basis of ease of access and the rigidity of the workpiece during the machining operations. As a general heuristic the spread of the attachment points should be as large as is practicable.

Bolt positioning

If bolting is the selected method of attachment, this module identifies the tapped holes in the platen most suitable for blank attachment. The hole locations are chosen with a view to minimizing the blank size together with the provision of maximum rigidity during the machining process.

In machining, detachable waste can be a serious safety hazard. This is particularly so where the component is machined from both sides or recesses occur in the outer profile. This module aims to identify such occurrences before making a decision to between bolting or area clearance of the offending waste.

Dowelling

Double-sided components demand accurate location of the two machining sequences. This is achieved by the insertion of two dowels in an area of waste material which locate in reamed holes in the platen. This module contains the rules for the selection of the dowel position.

The machining strategy is based on using the largest possible standard cutter to remove the bulk of the metal before reverting, if necessary, to smaller cutters for finishing or to provide the finer details. For double-sided components the second face is machined after turn-over.

The output from the AMP is in a neutral data format which is post-processed to the chosen machine tool. The post-processors used incorporate the necessary technological knowledge to select the feeds and speeds appropriate to the to the selected material, cutters and machine tool.

Verification and planning documentation

A range of check plots of component geometry and cutter centre paths are output (Fig. 13.2). Both line drawings and ink-jet coloured plots are used as aids to verification.

In addition AMP produces a comprehensive set of planning documentation to control the various stages of the manufacturing process. The set includes:

- The master planning sheet detailing the component identification, material, group technology cell selected together with a list of the blank preparation and machining operations (Fig. 13.3).

- Plots showing the blank preparation details (Fig. 13.4) and the set-up procedures for each side to be machined, including dowelling locations for the second side.

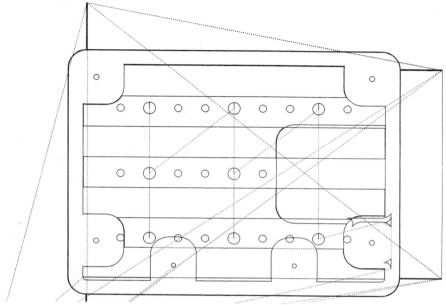

Fig. 13.2

- Operator instruction sheets giving machining datums, tool descriptions, minimum tool projections, drive tape sequences and the tape running time (Fig. 13.5).

- The part-off sheet showing the partly machined component in location on the platen and providing instructions on removing the bolts or clamps during the part-off sequence (Fig. 13.6).

Appraisal of AMP

AMP has clearly demonstrated the feasibility of developing a computer package to perform the functions of a part-programmer for the manufacture of 2½D parts from cuboid blanks. The program has a number of limitations including the following:

- Since the knowledge used has been encapsulated in a FORTRAN program rather than being stored as a separate body of rules, it is difficult to update and extend the system to cater for new knowledge or to deal with a wider range of components.

- The geometry of the parts is held in a relatively simple form, which although adequate for a wide range of 2½D components manufactured from blanks, is not suitable for extension to preformed components, such as castings and forgings.

MACHINE SHOP COPY

OP NO.	OPERATION DETAILS	TAPE/TOOL DESCRIPTION
	MASTER PLANNING CONTINUED	
10	CUT BLANKS TO SIZE	NON FERROUS SAW
	9.125" X 7.125" TOL +-1/8"	
20	FLYCUT BOTH SIDES TO THICKNESS	MASERATI/CINCI 2V
	1.580" +-.003"	
	FLAT AND PARALLEL TO .003 F.I.M.	
30	DRILL BLANK AS PS 3	BLANK DRILL
40	SET UP AS PS 4	MATCHMAKER CNC
	MACHINE AS PS 5	11259/101BT10
50	SET UP AS PS 6	MATCHMAKER CNC
	MACHINE AS PS 7	11259/101BT20
60	DEBURR	DEBURRING BENCH
70	INSPECT OPERATIONS 10 TO 60	INSPECTION DEPT
80	TAP ALL HOLE(S) AS DRG	ESSEX TAPPER 2BA
90	EXAM ALL OPERATIONS COMPLETE	INSPECTION DEPT
	PS 11259/101B SHEET 2	ISSUE 1

Fig. 13.3

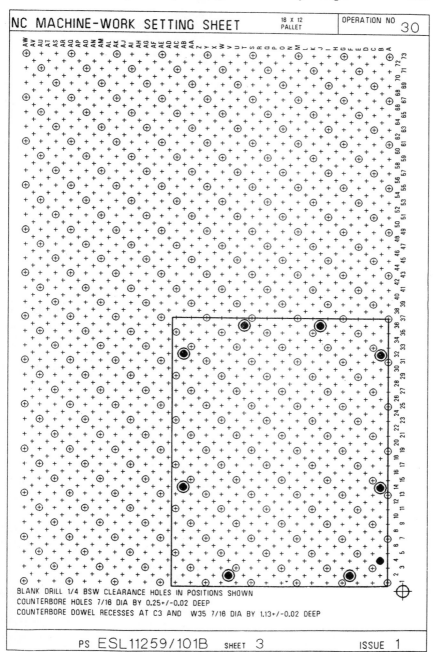

Fig. 13.4

Geometric Reasoning

```
JOB NUMBER : ESL11259/1018T20 ISS0          CREATED : 4-JUN-1985 (12:11)
OPERATION NO.  50                           MACHINE : POSIDATA CNC 2800
MATERIAL : ALUMINIUM                        PLANNER : O ROBERTSON
```

XY DATUM : STANDARD MACHINE SET POINT										
Z DATUM : 3 IN ABOVE PLATTEN/BACKING SHEET										
MINIMUM PACKING THICKNESS : 0.067										
MAXIMUM Z MOVEMENT : 2.993 X LIMITS : 0.000 9.300 Y LIMITS 0.000 7.625										
ESTIMATED RUN TIME : 46.39 MINUTES										

TLC NO.	DIA NO.	BLOCK NO.	TOOL DESCRIPTION	MIN.PROJ. INS.	SPEED R.P.M.	TIME MINS.	NO.OF HOLES	HOLE REF.	LOW LIMIT	HIGH LIMIT
1	-	5	0.750 IN. SLOT DRILL	1.50	2292	13.11				
2	-	2305	0.500 IN. SLOT DRILL	1.13	3438	2.34				
START OF DRILLED HOLES SEGMENT										
3	-	3060	NO.1 CENTRE DRILL		3600	2.99	22			
4	-	3550	4.00 MM. DRILL		3600	1.41	16	C	0.1567	0.1626
5	-	3920	2.80 MM. DRILL		3600	1.50	4	B	0.1094	0.1154
6	-	4290	2.30 MM. DRILL		3600	0.92	2	A	0.0898	0.0957
1	-	4530	0.750 IN. SLOT DRILL	1.50	2292	4.77				
7	-	4990	0.750 IN. LONG SERIES SLOT DRILL	3.22	2292	6.84				
RECLAMP AS PER PLANNING SHEET 8										
1	-	5765	0.750 IN. SLOT DRILL	1.50	2292	2.83				
7	-	6110	0.750 IN. LONG SERIES SLOT DRILL	3.22	2292	7.69				

Fig. 13.5

- Since there is no in-built explanation facility within AMP, it is difficult to obtain meaningful feedback from users on improvements to be embodied as a result of experience.

- It is difficult to assess the effect of aspects of the geometry, for instance thin walls and detachable waste, on the cutting and clamping strategy.

- The success achieved and the identifiable potential for improvement led to the decision to proceed to the second phase of the work.

An intelligent knowledge-based system approach

In an attempt to overcome the major limitations of AMP, the second phase of the work is concentrating on the creation of a part-programming knowledge-based system which will link directly to and will communicate with a solid modeller. The knowledge about the existing part-programming process is being reformulated as a separate set of rules in a form which is understandable to the user. In the light of experience gained, the rule base will be updated and extended. An important feature of the new system will be an explanation facility which, when challenged, will explain to the user the rules used to reach a decision. This is essential for the refinement of the rule base.

The solid modeller provides a well structured and easily modifiable representation of the geometry of the workpiece and other associated hardware, including jigs, fixtures and cutting tools. Current software available can be used to modify the models as required. A major part of the project will concentrate on creating an effective method of

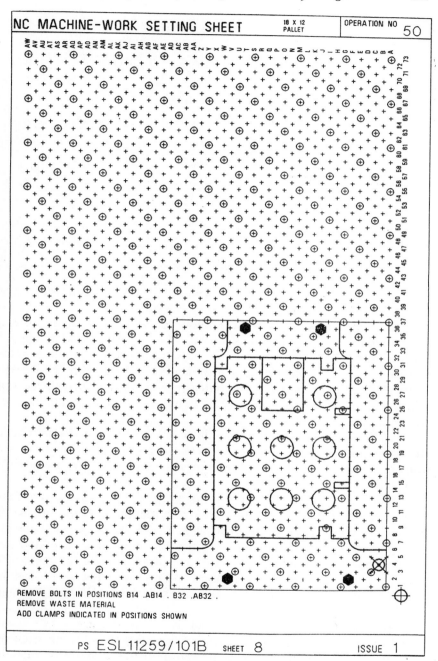

Fig. 13.6

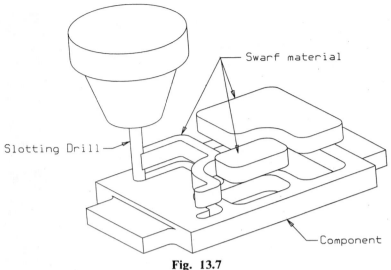

Swarf material

Slotting Drill

Component

Fig. 13.7

interrogating the model for information relating to geometric relationships and features.

Fig. 13.7 shows an example of how the modeller would subtract cleared columns from the blank form. The strategy for updating the part-machined model has not been fully established. A compromise must be struck between holding a model compatible with the current stage of machining and the computer power required to achieve this objective. Initial investigations would point to updating the model prior to any interrogation requirement.

Detachable waste parts are a serious safety hazard. Their creation during the machine process can be detected by the modeller, as a new body will be produced if a part becomes detached from the parent body. This is illustrated in Fig. 13.8.

One of the most difficult tasks in this project is the extraction of data from the modeller. A hierarchical interrogation of the model is envisaged, starting with the gross shape and working progressively through specific features down to the detailed geometry. The modeller must provide the necessary strategy validation as well as geometric details for cutter path determination. Currently, the extraction of 'implicit' knowledge from the model can only be undertaken through user interaction.

The first objective of the project is to reconstruct AMP using knowledge-based system techniques. Currently, the existing decision making processes of AMP are being reformulated as a set of rules in Prolog. Because of the need to manipulate three-dimensional data, this system will interface directly with a solid modeller.

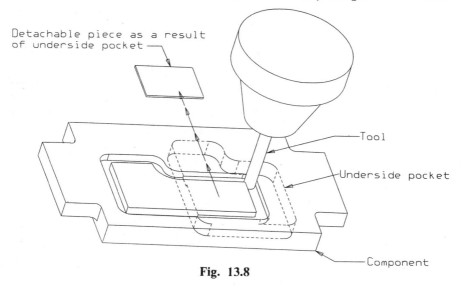

Detachable piece as a result
of underside pocket

Tool

Underside pocket

Component

Fig. 13.8

For this purpose the necessary links to the modeller are being defined and developed to provide the facilities required by the knowledge-based system. This will enable the system, in developing its manufacturing strategy for the component, to query the modeller as required. This process will start with a high degree of user interaction with the aim of moving towards increased automation as the project progresses.

A knowledge-based system to aid assembly design

Approach

In designing an assembly a designer is inevitably aware of a number of relationships which exist between the different components of the assembly. If a CAD system is used to record the design, these relationships are captured implicitly in the model. However if the model needs to be changed, it is not always easy to keep track of the consequential changes that must be made to other components in the assembly.

Some systems go part way towards solving this problem by providing parametric design aids for the design of 'families of parts'. These design aids are normally implemented as computer programs in some standard programming language. Programs such as these are generally large and complex and very difficult to adapt if the design rules change.

Ideally one would like a flexible system in which knowledge about relationships between components and the dimensions of individual components is stored as a set of rules which may be easily comprehended and modified as and when required. For this reason a knowledge-based system has been developed as an aid to the designer responsible for the design of assemblies. In particular it is intended for use in the design of families of parts where the gain in productivity which can be achieved, is likely to be significant. The system maintains a base of knowledge comprising factual information about components of the assembly and their inter-relationships together with a set of design rules. It provides two basic modes of operation:

- **Edit mode**, in which rules and facts about a design may be entered or updated.

- **Design mode**, in which, by assigning a set of values to the parameters of an assembly, an instance of the assembly may be generated which satisfied all the design constraints. This may be used to follow the effects of changes in the design parameters as well as to investigate the effects produced by changes in the design rules.

Knowledge representation

The information/knowledge which is stored in the knowledge base is declared as a sequence of modules. Each module is one of two types:

Part modules

Each part module declares the attributes associated with a component. These include the name of the component, its shape, dimensions, etc. Some of these might be specified by the user, others generated by rules. For example, an elementary part module with user defined attributes might be:

    Part module : cylinder
    att (len)
    att (diam) .

A complex component may be built up from more elementary ones using the system options 'apartis' (which defines one module to be part of another) and 'atypeof' (which defines a part to be of a type previously declared). An example of this is given in the definition of a stepped shaft which is made up of a set of three cylinders, for which a definition might be:

    Part module : steppedshaft
    apartis (end1, cylinder)
    apartis (end2, cylinder)
    apartis (middle, cylinder).

The values of attributes may be determined by means of assignment rules in the body of the module definition, for instance:

```
Part module : steppedshaft
apartis (end1, cylinder)
apartis (end2, cylinder)
apartis (middle, cylinder)
end1 : len : = 1.5*end1 : diam
end2 : len : = end1 : len
end2 : diam : = end1 : diam.
```

The value which an attribute might take may also be limited by constraint rules, such as:

```
Part module : steppedshaft
apartis (end1, cylinder)
apartis (end2, cylinder)
apartis (middle, cylinder)
end1 : len : = 1.5*end1 : diam
end2 : len : = end1 : len
end2 : diam : = end1 : diam
middle : diam > 1.5*end1 : diam.
```

Or one may define a gearwheel as follows:

```
Part module : gearwheel
apartis (blank, cylinder)
apartis (hole, cylinder)
hole : len : = blank : len
att (noofteeth)
att (toothdepth)
blank : len > = 1.2*hole : diam
blank : diam > 3*hole : diam.
```

Relation modules

A relation module defines a type of relationship which may exist between components. For example, the relationship 'isforcefitin' might be defined as

```
Relation module : isforcefitin
parameter1 : shaft type steppedshaft
parameter2 : wheel type gearwheel
wheel : hole : diam : =0.999*shaft : middle : diam.
```

Such a relationship may be involved from within a part definition. The most obvious way of doing this is in a production rule. For example:

Part module : gearassembly
apartis (shaft, steppedshaft)
apartis (gwheel, gearwheel)
shaft : end1 : diam : = torque/100
torque > 100 ⇒ isforcefitin (shaft, gwheel).

The user interface

The user may interact with the system via an editor which enables the insertion, deletion and update of facts and rules in the knowledge base. The editor is a menu-driven sub-system which operates at two levels, the module level and the rule level. At the module level whole modules may be inserted, deleted or renamed; at the rule level facts and rules within a module may be inserted, deleted or changed. The editor checks the syntax of the rules entered; it also performs certain basic checks on the operations performed and warns the user against such problems as insertion of a rule that gives a value to an attribute which is already defined in another rule, deletion of a module that is referenced by another module, etc.

Once a complete assembly has been created the user may generate instances of the assembly by entering the design subsystem. This part prompts the user for values for the design parameters and then uses the design rules to produce a complete design description.

13.5 Conclusions

In order to build more powerful knowledge-based systems to aid in various aspects of the processes involved in the design and manufacture of products, it will be necessary to develop more advanced three-dimensional modelling systems. Besides the pure geometric description of the shape of the artefact, such systems will need to store various types of knowledge concerning the features, functions and relationships relevant to it. It is also desirable to provide an interference capability to enable the modeller to use this knowledge in order to answer queries at a higher level than is currently possible. The two examples which are described in this paper illustrate different aspects of this need.

Acknowledgements

The authors would like to thank J.Y. Alexander and K. Edmonstone of Ferranti Defence Systems Ltd., who undertook much of the original work on the AMP and also the ACME Directorate, FDSL and Ferranti Infographics Ltd. for supporting the current research programme. Thanks are also due to I.E. Aitchison and G.A.R. Wilkie of Heriot-Watt University who developed the knowledge-based system for the design of engineering assemblies.

References

1.   J.Y. Alexander, 'The automatic generation of control and planning data for the manufacture of 2½D milled components', BNCS Annual Technical Conference, Brighton, 1982.

2.   G. Bosch, 'NC programming developments', BNCS Annual Technical Conference, Wembley, 1981.

3.   B.J. Davis, I.L. Darbyshire, A.J. Wright and M.W. Park, 'The integration of process planning with CADCAM including the use of expert systems', Proceedings of an International Conference on Computer-Aided Production Engineering, Edinburgh, MEP (35-40), 1986.

4.   H. Stolenkamp, W.J. Oudolf, H.J.J. Kals, 'Cubic—a highly automated system for programming NC machining centres', SME Technical Paper Series MS n MS79-167 (22) 1979.

5.   A.H. Van't Erve and H.J.J. Kals, 'XPLANE: a knowledge base driven process planning expert system', Proceedings of an International Conference on Computer-Aided Production Engineering, Edinburgh, MEP (41-46), 1986.

Discussion

(Presented by Murray and M. Williams).

Middleditch: First of all, a comment: I agree with you thoroughly about this business of part-program proving and looking at plots of cutter centre-lines and so on. We've tried lots of different ways of doing it, they're all useless, and the only way I have found practicable was to watch a volume model transform itself from a blank into a component on the screen, and that's a very good way to know that you're doing the right job, and that's the one we've locked into.

Another thing about this 2½D question: the handwheel that Bowyer showed us (Fig. 1.2) was made by 2½D machining. Nobody in his right mind would look at that and say that it is a 2½D component. I wonder if its possible to have a definition of 2½D; maybe we shouldn't even try. Whatever it is, someone will come up with a counter-example.

Murray: That's right; and it depends on the geometry of the form tools that you want to use. That handwheel was made with a ball-nosed cutter: a sphere traversing the surface?

Bowyer: Yes.

Prior: I wonder if you were going to infer forward from the algebraic constraints that you derived?

M. Williams: No!

Martin: Can I just say that I feel that the people who want to do toolpath verification are showing rather a lack of confidence in their modellers, if they feel that such verification steps are necessary.

Bowyer: Have you ever sat next to a machine tool running a part-program for the first and most interesting time?

Middleditch: I agree with Martin entirely. We don't bother with toolpath verification; we generate tapes that are correct the first time!

Dodsworth: Firstly, a comment on verification: in our experiments we generate the tool motions as solids, subtracted them from the stock, and then compared them with the initial solid. That comparison generates some solid models which are what some people call, 'pathological'. (Although, in my experience of computational geometry, the definition of 'pathological' is something that happens about 99% of the time!) And in fact they do pose rather special constraints on solid modellers. Because they have lots and lots of coincidence, although the thing that you hope to end up with is nothing (the empty set) so they are interesting in their own right. But you were talking about wall thickness; you wanted to use a solid modeller rather than just a list of z-depths and profiles to determine wall thickness. My feeling is that your first representation was probably much more useful. I think it is probably easier doing it from a list of profiles than from a solid model.

Murray: Yes. In some ways, what we were saying, was, 'Developers of solid models, can you help us?'. These are the questions that we are posing. Now, we can do it to a certain extent with our 3D profiles, and those Z-depths, but its a very special case, and we haven't got a general algorithm developed for that but, as you say, we can do it from the 2D profiles.

Williams: But it does mean more work, because you have to create the 2D profiles, rather than work direct from the model.

Murray: 2D profiles have to be created anyway.

Dodsworth: 2D profiles are in any case the best input to a solid modeller to create a 2½D toolpath.

Middleditch: I was intrigued with your process plan. When I looked into process planning some time ago, I tried to determine what a process planner does, and after a week of sitting with him, I didn't know.

Sabin: He writes a planning sheet!

Middleditch: Yes, that's right, but he writes things like 'When you've got two holes, drilled one into the other, you do the little one first': little things like that. How do you build that into your program, because there are millions of those little things.

Murray: Well, these are the type of problems that at present we have problems with, with our FORTRAN tool path planner. There are two things that have been against the FORTRAN version of our program. Little things like that, and some big things, like trades unions!

Middleditch: So those little things are made explicit—you don't have general rules?

Murray: No. We keep coming across these little things with the old version of our program, particularly if you have things which are possibly not true $2\frac{1}{2}$D, where you have features coming in from the sides that might interfere.

M. Williams: That's precisely why you want a knowledge base; because you've got little rules like that and, in some cases, conflicting rules. Which boils down to the sort of things discussed in Mill's paper [6]—you need a costing association. In some cases there may be safety aspects. There isn't a simple strategy to tie these up; the approach we have taken is to have rules and meta-rules to decide between rules.

Middleditch: But what annoys me is, it's all physics. Why can't we have some general way to establish all these little rules? I couldn't come up with one, but there ought to be one—it's all physics!

Mill: The work that we're trying to do is to develop a process-planning shell, if you like, into which people can fit their own rules: for the simple reason that we couldn't actually write any general rules.   The example about the component fitting on to the machine is one. We tried to say that this would be universal to every company, but it turned out to be wrong. The only things that are universal in process planning are the laws of physics, so the approach that we have taken is to allow people to write their own rules. I'm not too sure about how many rules were there actually be in a system; it worries me; and if as you say the answer is 'millions', then you've got to start wondering about the feasibility of using a knowledge-based process planning system. If it's of the order of thousands then we may get away with it.

Sabin: In the typical company it's between six and twelve feet of shelf space, isn't it?

M. Williams: I don't believe that a process planner is actually dealing with millions of rules. This is what people might have thought in terms of medical expertise until people actually started building medical expert systems; then they found that it did boil down to a relatively small number of rules. The rules are not purely physical. You are weighing physics against economic and other factors.

Murray: The other thing that is also interesting is that exactly the same simple path can be given to three or four planners, and they'll all plan it a different way, and they'll all argue that their way is the best. There must be one way better than the others, but it's actually quite difficult to decide on weighting factors to come out with a figure of merit for each way; and in many cases it doesn't matter which way you do it.

Braid: What's the main aim? Is your aim to make process plans better, or avoid mistakes, or make a part faster?

Murray: We want to be able to make process plans better, but that is not that important. The aim is really to cut down the wait time and increase reliability. We can take a piece of geometry, and we put it in. It says either that we can or cannot deal with it. If we can deal with it in the 'black box', then what comes out will be correct. We hope that we can put the blank on the machine table, and stand beside without being in fear and trembling, while it cuts a small batch of components.

Middleditch: Just like the definition which eludes me, I don't know what a process plan is. I'd be interested to know if anyone has a definition of one.

Murray: I think that there's a definition in the paper of what we thought it was.

M. Williams: Surely Fig. 13.3 sums it up.

Middleditch: I'm looking for a more precise definition that eliminates all those things that aren't in it and includes all those that are, and every time I see a process plan, and ask somebody this question, I get a different answer. My contention is that there isn't a definition, and I'd like somebody to tell me if there is.

D. Williams: I think that you can take a pragmatic just-in-time definition. 'A process plan is the description of the process that gets the component out of stock, into the warehouse and on the wagon to the customer!' Everyone has been describing chip-making process plans, we haven't heard anything about other processes.

Murray: Yes, there are certainly a lot of downstream processes that also need planning.

Additional reference

6.   P. Husbands, F. Mill and S. Warrington, 'Part representation in process planning', in *Geometric Reasoning* (Proceedings of a conference held at the IBM UK Scientific Centre, Winchester, December 1986), OUP, 1989.

14 The context of geometric reasoning in manufacturing cells

R. K. STOBART and D. J. WILLIAMS

14.1 Introduction

Context

This paper discusses the application of geometric reasoning to the planning and execution of operations in a manufacturing cell. The manufacturing cell is the fundamental building block of manufacturing systems and, like a biological cell, is the smallest unit of sustained productive activity [32]. The cell may be concerned with metal cutting, assembly or other manufacturing operations. The three aspects of cell operation we will consider are:

- Off-line planning of cell operations which may include the layout of the cell.
- Planning for operation of the cell.
- Error detection and recovery in the cell.

Each aspect has a geometric content. Where geometric reasoning needs to be applied in particular is to the first and third of these aspects. Normal cell operation is the result of successful planning and recovery from errors.

Drives behind the work

There are two main drives behind the work:

- Cell design and planning.
- Error recovery during operation of the cell.

During cell operation, errors may occur due to unforeseen circumstances; automatic error recovery is essential to keep the cell in

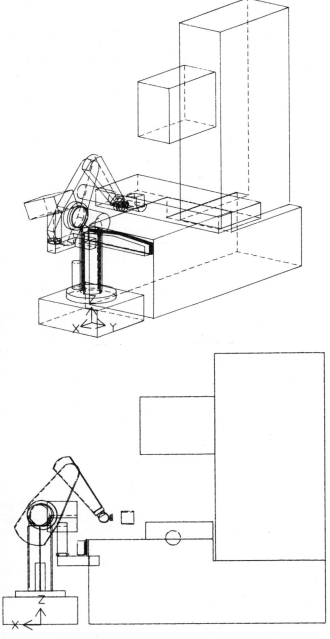

Fig. 14.1 (above) Oblique view of a milling machine being supplied with parts by a robot, produced by the BUILD solid modelling system.

Fig. 14.2 (below) Orthographic view of the scene from Fig. 14.1.

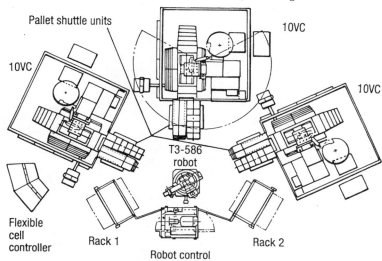

Fig. 14.3 Plan of the entire machining cell, with three milling machines fed by a single robot.

production and avoid costly down-time. Most error-recovery operations will involve movement of the cell components and hence there is a strong geometric element in the error-recovery process. We are interested primarily in error recovery which is interactive, with sensors in the cell, and is goal-directed in the sense that some level of production can be resumed.

Much geometric information is available at the planning stage and a link is required between off-line planning and on-line error recovery.

Examples

- Robot handling into a machine (Figs. 14.1 – 14.3): The robot 'serves' the machine tool by loading and unloading work. The cell operation is coordinated with the supply of components from other processes. The robot may also assist the machine tool should the tooling or fixing need changing.

- Cooperating robots (Figs. 14.4 – 14.6): Two or more robots may be employed in an assembly operation to reduce cycle time or to use the capabilities of two different forms of robot. This technique has been reported by industrial users [24].

- Robot working with a hoist (Fig. 14.7): The robot may need to handle heavy parts where a heavy-duty robot is too expensive. Here

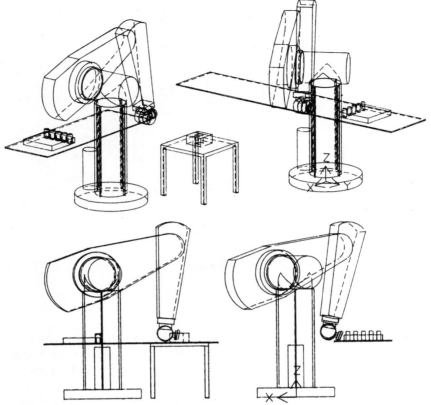

Fig. 14.4 (above) Oblique view of an assembly cell, produced by BUILD.
Fig. 14.5 (below) Orthographic view of the scene from Fig. 14.4.

a balancing lift is required which the robot drives in the same way as a human operator.

State of the art

Commercial developments in this field have been proceeding for some time. In this section we will briefly review what is the 'state of the art' in terms of commercially-available software systems.

Cell simulation

In the last five years growing interest in off-line programming of robots has led to the development of a number of robot simulation packages. While the more recently-announced packages have some quite

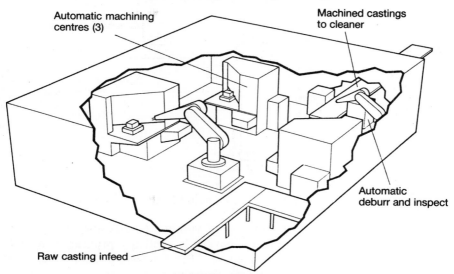

Automatic machining centres (3)

Machined castings to cleaner

Automatic deburr and inspect

Raw casting infeed

Fig. 14.6 IBM 9335 manufacturing facility.

sophisticated facilities [2,19,37], the basic philosophy of the techniques is much the same and can be summarized as follows:

- Robot models and their kinematics are read from a 'library' to which the user can add as necessary.

- The robot may be driven between points and along paths and its movement observed and analysed using a graphic display.

- The robot may be programmed in a high-level language which is later translated into the robot's target language; alternatively the commands used during the simulation may themselves be post-processed into the robot's own language.

- The robot's motion may be coordinated with other equipment in the simulated environment; this coordination is determined without taking account of statistical variations in the timing of different processes.

- Collisions are detected graphically or, in a few cases, by analysing the geometries of object models in close approach.

Cell programming and error recovery

Most cell programming is done in a 'traditional' way by programming the cell components in their respective target languages. This can involve a wide range of languages (from Pascal through explicit robot

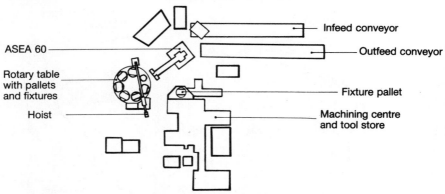

Fig. 14.7 Cell with hoist.

languages to relay ladder logic) in large volumes (> 10k lines of code for
a typical assembly cell). Error recovery is explicitly programmed and
forms the bulk of cell-control code. Error-recovery techniques are
necessarily quite limited and are generally unable to draw upon a world
model to assist in making a recovery plan.

14.2 Cell planning and control

Aims

The aim of cell planning is to reduce the time required for the design
and build cycle of a new manufacturing cell. Planning and modelling
will also be required when the cell function is changed so that it will be
ready to start production of the new items as soon as the previous batch
has been completed.

The major objective of planning is the production of the required
items. However, there will be subsidiary objectives such as the meeting
of a certain maximum cycle time or the occupation of a certain
maximum area of floorspace. A more esoteric objective may be that the
robot fulfils a certain accuracy requirement through the disposition of
sensors and the siting of the robot.

A further consideration is whether any part of the cell can readily
be replaced as part of the repair operation. This requires that the plan
is formulated in such terms that a new component can easily be added
and brought into production. For a robot this may be difficult unless
the plan has been configured with the robot's inherent inaccuracy in
mind.

Planning

A background to planning

Planning could be defined as 'the successive refinement of a high-level goal to produce a sequence of atomic operations'. Our goal in manufacturing is to produce an item. However, in arriving at the plan we must meet a wide variety of objectives.

The plan for a manufacturing cell will consist of a large number of atomic operations involving several different devices. For example, in a global plan for a machining cell there will be operations for the robot, the machine tool and the cell controller. We are particularly interested in the role of a robot in a machining cell; the robot's instructions will consist of movements of items around the cell and synchronization with other devices. Therefore the robot must handle sub-plans which have two main components:

- **Path planning**, which involves the planning of an efficient collision-free path along a trajectory or between points.

- **Grasp planning**, which involves planning the operations required in order to grasp an object safely.

We are primarily interested in path planning, which may involve compliant motion. We assume that there are efficient heuristics for grasp planning where the objects are prismatic [17].

Another minor area of interest is in planning for sensing, in which the basis for a sensed event is planned in advance [30]. For example, the likely configuration and position of a component may be pre-computed for vision processing to ease the on-line computation requirement [1].

To produce a robust plan 'reinforcements' must be provided. Carroll and Williams [7] mention three forms of reinforcement:

- **Redundancy** implies that, if a plan component fails, there are other ways of achieving the objective. For example, it may take the form of multiple routes for a robot between two points so that is one is blocked the other may be used.

- **Contingency plans** may be included which anticipate certain failures. They are brought into operation when the main plan fails. For example, if a peg-in-hole insertion fails we may wish to try searching for the hole.

- **Deferral** is a technique in which the final resolution of the plan is deferred until sensor information is available at run-time. This is the technique used by Yin in the vision extension to RAPT [45].

The plan includes a section in which certain variables are left unresolved and are finally instantiated at run time when vision data can be processed.

Planning and scheduling

Planning in the context of manufacturing cells has been reviewed elsewhere [7]. The aim of planning is to produce a specification for the operation of the cell which meets the performance objective. Planning is complete before run-time and contingency needs to be built in to allow the plan to accommodate unforeseen circumstances.

In scheduled operation the cell functions are organized at run-time in response to sensor readings which indicate the state of the cell. An example given by Fox and Kempf [14] is that of a robot conducting an assembly in which the parts are presented in a bin. The robot collects parts from the bin in turn and schedules the assembly on the basis of which parts are selected. Fox and Kempf's emphasis is on the use of 'opportunism' in scheduling to make use of particular orderings of parts to make the assembly operation in some way easier.

Planning and scheduling have different geometric requirements. In planning, a series of geometric operations are specified and method are used to make modifications at run-time (deferral). Scheduling consists of generating plans for a series of atomic operations which can be rearranged in time to suit the run-time conditions. The geometrical operations can be planned for each of the atomic tasks: for example, the robot movements required to assemble two components. However, a geometrical 'framework' needs to be provided for the scheduling operations. We will explore more aspects of such a framework later when error recovery is discussed.

Path-planning techniques

In this section each of the main planning techniques will be reviewed. The salient work on path planning was done by Lozano-Pérez and Wesley [23], who formulated the configuration-space approach to the problem.

Configuration space has a dimension equal to the number of degrees of freedom of the manipulator. A point in configuration space is then a representation of one particular manipulator configuration. An illegal configuration due to an intersection of the manipulator with workspace objects produces configuration-space obstacles which can have a very complex structure [16]. The notion of configuration space has provided a basis for some useful results. For example, if at the start and end of

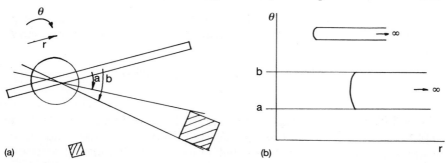

Fig. 14.8 A cylindrical manipulator (a) is surrounded with obstacles. In (b), the obstacles are drawn in configuration space.

a motion two objects are in contact then a motion exists in which the objects stay in contact throughout the motion. In some of the literature the problem of finding a collision-free path is referred to as the 'findpath' problem. This is a general term but is most often applied to the 'free-space' techniques discussed shortly. A simple example illustrating configuration space is given in Fig. 14.8.

Algorithmic techniques

Algorithmic techniques have recently received a great deal of attention [41]. They are based on the theory of algebraic topology and several different methods have been developed for planning simple robot motions. Schwartz and Sharir [35] show that the motion-planning problem can be solved in a time which is a polynomial in the number of algebraic constraints defining the free space but doubly exponential in the number of degrees of freedom of the robot.

In the projection method some of the degrees of freedom of the problem are fixed. The planning problem is then solved for the reduced-dimension case. The result of the sub-problem is a representation of the free configuration space which changes at critical values of the fixed degrees of freedom. The entire free space is partitioned into connected cells, and the full problem may be solved by path searching techniques. The method has been applied to the 'piano movers' problem' [33,34,36].

In the retraction method the configuration space is retracted on to a lower-dimensional sub-space. If two system positions lie in a connected component of the sub-space then they will also be in a connected component of the configuration space. Several applications of the method have been reported [27,28,29]. They deal with path planning in very simple environments.

In the present context of moving objects or changing robot configurations the planning problem becomes substantially more difficult.

Heuristic techniques

The simplest of the heuristic techniques is 'hypothesize and test' in which a candidate motion is selected between initial and final configurations. At certain selected points in the planned trajectory, collisions are investigated; if one occurs the trajectory is modified by taking into account the form of the collision. This is by far the most common technique in practical use today, primarily because of its simplicity. While collision-detection methods are well understood, path modification presents difficulties. In an uncrowded environment simple heuristic techniques can be applied, but as the environment becomes progressively more crowded these techniques break down. Successful use of the technique requires a high level of human intervention which is desirable for cell planning but renders the technique useless for automatic path planning. (Widdoes [44] reports some early work on heuristic methods.)

'Semi-heuristic' techniques

We can define a class of 'semi-heuristic' techniques which in themselves are not generally able to solve the 'findpath' problem but are well suited to adjusting a proposed path so that it is collision free. Khatib's potential-field approach [20] typifies these techniques. His approach is to postulate potential fields around objects in the workplace which present a collision hazard. The target point for a moving object is considered to be attached to the robot by a spring. The potential forces are then balanced by an attractive force which 'pulls' the robot to its target position. The resultant forces can be resolved into joint torques. The effect of the potential method on a simple trajectory is illustrated in Fig. 14.9.

Khatib has demonstrated the technique for real-time obstacle avoidance and its application to the off-line case is straightforward with some modifications. These techniques seem to offer a substantial improvement over pure hypothesize-and-test methods and should be seen as the next step forward before the full development of 'free-space' techniques.

We can suggest how such a class of techniques might work.

- A path is suggested by a programmer which may be based on a sequence of points selected from the cell model. This implies a

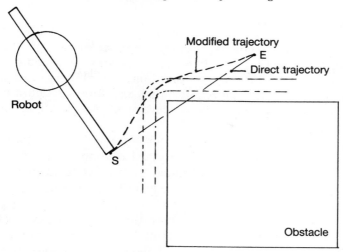

Fig. 14.9 Path correction by the potential method.

requirement for a range of interpolation techniques appropriate to the problem. While cubic splines have been suggested because of their continuity properties [22], better techniques are required which have less tendency to oscillate.

- The path is tested by driving the robot model along it. Here we need an efficient collision-detection technique supplemented by minimum-distance monitoring. It is not adequate merely to 'expand' obstacles to compensate for uncertainty in the robot path; instead we should have an explicit minimum-distance criterion. Another problem is the criterion for choosing points along the path. This problem has been addressed by Cameron [6], but Stobart [39] has suggested a technique which permits a 'time-history' analysis of a trajectory segment which reports collisions even when the sampled points are widely spaced and the actual point of collision is missed.

- If a collision is reported the trajectory is modified using a variant on the potential-field method. As a final check the trajectory can be displayed and rechecked for collisions.

While this method has not yet been developed the algorithms could be made quite efficient on current computer architectures. Furthermore the use of such methods in real time is plausible so that the application to real-time error recovery can be considered.

'Free-space' techniques

In the free-space methods, explicit representations of the robot configurations which are free of collisions are built. The obstacle-avoidance problem is then one of finding within these representations a path connecting initial and final configurations. A number of methods have been developed which differ in their representation of free space.

Udupa [41] describes a technique for representing free space for the Stanford arm which consists of a vertical column, a horizontal boom and a wrist. The arm is approximated by two cylinders corresponding to the column and the boom respectively and the free space is computed for the boom cylinder. The free space is represented in configuration space as cuboids formed by the first three joint angles. The safe trajectory for the manipulator is obtained by recursively modifying the straight-line path between initial and final configurations until it lies completely within the manipulator free space.

Brooks and Lozano-Pérez [5] describe a method of planning the motion of an object in two dimensions with rotations permitted. With three degrees of freedom the configuration space is three-dimensional and is divided into cuboids. The cuboids are each given a status which can be:

- Empty: if the cuboid does not intersect a configuration obstacle.
- Full: if the interior of the cuboid everywhere intersects one or more configuration obstacles.
- Mixed: if there are interior points inside and outside configuration obstacles.

The path is found by searching for a connected set of empty cuboids. If such a path cannot be found then mixed cuboids must be divided up into smaller full, mixed and empty cuboids until either a path is found or failure is indicated. Brooks' work has been extended to the planning of pick-and-place operations using a robot with six degrees of freedom [4]. He simplifies the problem by separating the motion into two parts: that of the lower part of the robot and the payload, and that of the upper arm. Free space is represented in two ways. For the planning of payload motion, freeways are defined in Cartesian space. The upper arm motion is characterized by the first two joint angles which may be more conveniently represented in configuration space. The choice of cuboid objects in the workplace ensures that they too can be represented in the configuration space. The path-planning problem consists of finding valid motions for the upper arm, then valid motions for the payload. Finally the selected motion is checked for the collision of the forearm with an obstacle. Brooks suggests that the algorithm can be improved, although it runs in quite acceptable times. While there is still a query over the simplification the algorithm requires, it could be used for practical path-planning work.

Uncertainty in the model

We can identify several sources of uncertainty in a cell model which may result from false or inadequate assumptions about the behaviour of cell components. Major sources of error are:

- Differences between a robot's real and planned trajectory.
- Variations in the position and orientation of items appearing as 'cell inputs'.
- Variations in the timing of external events affecting the cell.

We consider each error source in turn.

In most of the commercially available cell modelling programs, the robot trajectory is idealized as the programmed path. Some of the more sophisticated modelling systems [9] allow experimentally derived corrections to be applied, while at least one [43] uses the kinematic transformations actually used in the robot controller to reproduce some of the path-following and positioning errors. It is generally accepted that to include robot dynamics in a robot modelling system would substantially increase its complexity and cost, but at least one example [18] exists.

In the medium term, research is being aimed at faster and more accurate robots and the accuracy problem is likely to diminish. For pick-and-place robots the trend is towards lightness and flexibility so that between points the trajectory is likely to deviate significantly from the nominal while achieving the target points precisely.

For planning purposes we need to know the bound on the robot's movement which in general will be a function of joint position, speed and acceleration. This information could be computed and summarized as a set of non-linear functions from which values could be extracted as planning proceeds. A full dynamic simulator is therefore not necessary and the supplier of the planning system will do the necessary dynamic simulation to produce the 'summarizing functions' to be applied by the its user.

Uncertainty in the position, orientation and size of objects entering the cell has been the subject of recent research [13], in which uncertain geometry is treated in a probabilistic framework. By exploiting invariant topologies and Gaussian probability distributions, a general framework for handling uncertain geometry is established. An essential component of this approach is the modelling of the random processes which affect the cell, and we might apply established modelling techniques to the problem if we could define suitable experiments. Such model data can be combined using the geometric techniques defined in [13], which could be embedded within a geometric modelling program. By establishing uncertainties in the cell we are in a position to revise the cell design or to deduce the siting of sensors which maximize the positional information.

External events will occur on a timescale which is unlikely to be either consistent or predictable. It is the role of discrete event simulation to model the various interacting processes and to summarize their effect on plant performance parameters. The links between discrete event simulation and geometric planning are discussed below.

So far we have adopted a fairly passive approach to the problem of uncertainty. Daily [10] has demonstrated optimization techniques which allow a designer to predict the 'best' position of a robot with respect to a workpiece. The technique is based on the quality (condition number) of the manipulator's Jacobian matrix taken across the working area. The effect is that the robot works where its kinematic and control processing is best conditioned and thus a certain amount of positional and dynamic uncertainty is removed merely by positioning the robot.

Another way of coping with position and size uncertainties is through the use of compliant structures and compliant motion strategies. For example, in fettling, size variation is accommodated by compliant tooling, while in the well-known peg-in-hole insertion problem compliance is used to accommodate misalignment error in the insertion operation.

Uncertainties represent a significant problem but the subject is being tackled from a number of directions. What is required in the medium term is an integration of the various techniques to provide a unified assessment of uncertainty and a number of tools which can be used to reduce uncertainty either by sensor strategies or rearranging the cell.

Multi-level modelling

Computational resources limit the detail to which a cell model can be simulated. For example, we may approximate a sculptured surface in the work cell by a polyhedral surface to speed up geometric processes such as collision-checking and drawing. For best use of computer resources in planning, and later at run-time, a framework is required which handles a number of modelling levels. Let us consider a number of such 'levels':

- At the highest level of complexity we may wish to model the geometry of the cell as thoroughly as possible. The resulting graphics will probably be very attractive and provide a good basis for a presentation to senior management. However we can probably do a good job of planning the cell with carefully approximated models.

- At the stage of cell planning we initially wish to choose a robot and a 'rough' cell layout to provide the basis for later detailed design. We need to test the robot for reach, for the range of movements and for an estimate of cycle time. For these purposes a 'box' representation of the robot members will be adequate while items in the environment could be represented by enclosing boxes or wire frames.

- Next we wish to do more detailed checks of layout and cycle times and be certain that there are no severe collisions. Here a more detailed model of the robot is needed but the environment can still be fairly simple. A distinction needs to be made between gross movements and fine movements. During gross movements in relatively uncluttered space the robot model can be simplified, for example by reducing the gripper assembly to an enclosing cuboid; during fine movements the gripper must be well represented while the base and upper arm of the robot can be approximated because they will probably be relatively distant from other objects.

- In the last stage of cell planning the fine detail must be included so that detailed work can be planned and checked. For example, we might wish to do a detailed check of a machining operation on a complex surface. At the last stage we can introduce the full sculptured representation for path generation and collision checks.

In the medium term there needs to be research into 'adaptive' techniques of model representation which will be responsive to the degree of accuracy needed in the model. To remove the need to choose the degree of representation from the cell programmer, a degree of self-adaptation may be required which invokes a level of representation depending on the computed minimum distance.

Recalling free-space techniques, they need approximations of the cell environment which can readily be represented in configuration space. To support this path-planning technique we will need methods for approximating the environment for this purpose.

Integrating discrete event information

So far we have considered the manufacturing cell in isolation. This is a restricted view and to complete the modelling exercise interaction with the rest of the manufacturing plant must be considered.

The basis for the simulation of manufacturing plants is the analysis of discrete events [25,26]. Processes are modelled statistically and scheduled in a simulation structure; measurements are made as the simulation proceeds and statistics can be derived. The results of such a simulation are not immediately applicable to a geometric simulation of the manufacturing cell. However, the results can be processed to give the statistical distribution of events affecting the cell. We can then compute the probability of certain events which have a defined geometric effect.

If we consider the discrete event simulator as laying out a statistical basis for a cell operation we can submit the timing durations to a temporal reasoning program which will produce a critical path plan for the manufacturing operation. Supposing that we choose a probability level of 5% for this plan and find that the cycle time is too long, we should have a means of identifying which operations have high variability and overlapping these with other operations using scheduling techniques which will run in real time. To illustrate this consider the following simple example. A two-robot assembly cell must conduct two operations:

- A peg-in-hole insertion using compliant motion.
- A move to collect a new part from a conveyor.

The peg-in-hole insertion has a wide distribution of times while the move has a narrow distribution which depends on the dynamics of the supplying processes. If, in our initial plan, there is a high probability of collision because the two operations cause the robots to infringe each other's work space, then we would consider serial operation. However, if this causes a cycle time which is too long, we must consider a scheduling approach which resolves spatial conflict at run-time (Fig. 14.10).

Some achievements with BUILD

We have been implementing some aspects of cell modelling in the BUILD solid modelling program. Most of the work done to date has been reported elsewhere [38,39,40]. The cell-modelling tools include:

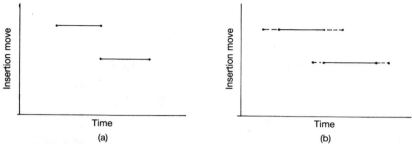

Fig. 14.10 In (a) the nominal times for two operations are shown. In (b) the effect of 'variability' is included. To avoid collisions, (a) make operations serial, *but then* 95% of time overlap is wasted, or (b) overlap operations and accept the need for real-time scheduling.

- Robot models which include inverse kinematics for path-following in Cartesian space. Collision and minimum distance checks based on 'boxed' versions of the robots and workplace objects.
- Time-history analysis of minimum distance trends.
- Multiple coordinated robot handling including machine tools and autonomous guided vehicles.
- Free-form surface modelling.
- Simple trajectory design and manipulation based on B-spline curves.

The cell models in Figs. 14.1, 14.2, 14.4 and 14.5 have been prepared using these BUILD facilities. Figs. 14.11 and 14.12 show the fixture being used in the machining cell illustrated in Figs. 14.1 and 14.2. A cuboid workpiece is located in the fixture and clamped using the lever. It is manoeuvred by the gripper, which is illustrated in Fig. 14.12 approaching the fixture. The level of detail shown in the figures is high and appropriate for collision detection in the close approaches involved in inserting and removing items from the fixture.

14.3 Error recovery

Motivation

The drive behind error recovery has several components. Primarily, the failure of a cell in a highly integrated factory has severe impact on overall performance, while for a cell operating in isolation, down-time is costly because of the degree of capital invested in the cell.

Manufacturing cells are likely to be operating in largely unmanned factories over two or three shifts, so an objective for error recovery is that it should be generally automatic and able to inform a maintenance computer of the nature of a problem so that plans for repair can be

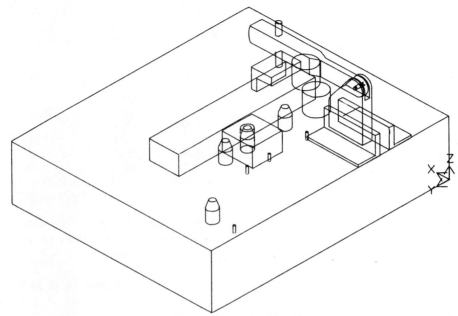

Fig. 14.11 BUILD model of a milling fixture.

formulated automatically. It is desirable that, if production can continue, it will. Error recovery should therefore be based on the sensed error, and will be an interactive process in which recovery is tried and evaluated before production is resumed.

An assumption underlying the error-recovery philosophy is that the environment is fairly chaotic and that errors are introduced by the environment as well as failures internal to the cell.

A basis

Error recovery has been the subject of much recent research [e.g. 12,15,21,31]. A framework for error recovery includes definitions of 'error' and 'failure' and a method of modelling the work cell.

A fruitful method of representing the work cell operation is as a state-space model [42]. The model has a number of 'states' which are fundamental parameters in the operation of the cell, for example, the voltage applied to one of the robot drive motors. The model accepts a number of inputs which will affect some or all of the states and produces a number of outputs. Restrictions on the number of sensors may mean that some of the states are not observable and also, depending on the cell's dynamics, some of the states may not be controllable.

During cell operation the state will be controlled by using the available inputs. The state has a wide definition and will include a

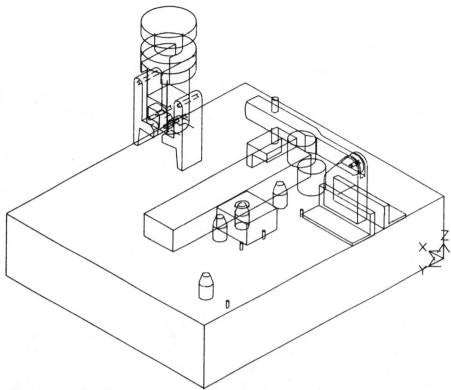

Fig. 14.12 The fixture of Fig. 14.11, showing the robot gripper approaching.

world model of the cell which describes the disposition of objects in the robot workplace. We can conceive of a state model in which the states include the position and orientation of the objects and details of their geometry.

A state model of a cell is shown in Fig. 14.13. The state of the system includes both geometric and non-geometric states of which only some are available at any time. Raw data from sensors is supplied to the cell controller and some is filtered to extract system states or simply to remove noise. Default data representing unmodelled aspects of the cell is also available. The cell controller acts upon the system states to produce known state transitions.

An 'error' has been defined [31] as a discrepancy between the measured state of the controlled system and the state of an internal world model where allowance is made for resolution in measurement. This definition of an error indicates that the main technical problem is in using sensor and other data to infer as many of the system states as possible.

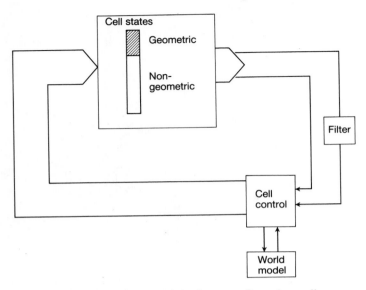

Fig. 14.13 State model of a manufacturing cell.

Having defined an error as a discrepancy between the measured state and the model state, the problem of automatic error recovery is seen to be the recovery from an error state to a known state from which a recovery can be made by known state transitions (Fig. 14.14). Certain states will not permit recovery and we can regard these as indicating failure. Such states will include those which would lead irreversible damage if an attempt was made to continue.

Automatic error recovery

Elements of an error-recovery technique

Cell operation is governed by a plan which will have been generated with both redundancy and contingencies [7]. There will also be assumptions inserted in the plan about what state should exist at certain points in the plan. The check on the assumed state (or **sensor profile**) [21] forms the basis for error detection. The form of recovery is based on whether the plan included contingency procedure for this particular error.

Modelling the cell

During execution of the plan an **execution monitor** must represent the state of the world using a model. Taking the state ideas a little further, the model must contain a wide variety of information including

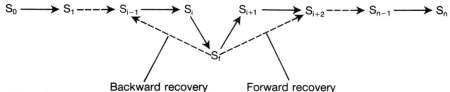

Fig. 14.14 Schematic of a task as a sequence of state transformations showing the occurrence of failure. $S_{i(i=0, n)}$ represents the ith state along the expected trajectory; S_j represents a failure state. The error-recovery problem is how to get to the final state from the failure state. Forward recovery may not be possible if the failure state has undone some of the previous transformations; backward recovery may then be necessary.

'logical' and geometric details. Logical information can be considered as either having some or no geometric significance. For example the presence of a part in a jig may be indicated as 'part present' which obviously has geometric significance; while the logical state of a relay in the cell electrical system has no direct geometric significance.

Geometric data about the fixed and moving parts of the cell needs to be stored in an efficient way which does not impede the execution monitor or the error-recovery process. It should ideally be derived from the cell plan already discussed. The geometric cell model should also absorb few computational resources until errors occur and recovery is invoked.

In a simple environment the cell geometric model can be readily updated because all relevant cell parameters are accessible: for example, robot joint angles or machine tool slide positions. However, in a complex environment, where parts entering the cell are of uncertain orientation or compliant motions often have unforeseen consequences, much more reliance must be placed on sensor data. If sensor data is continuously processed by an 'executive monitor' then the world model will be accurate and 'opportunistic' operation may be possible [14] at the expense of computational resources and elapsed time. A more efficient approach may be to process sensor data only when an error occurs so that updating the world model is the first step in the error-recovery process [11]. This approach has the drawback that a slowly developing error which could have been pre-empted is not, and a severe error or failure ultimately occurs.

The role of geometry in recovery

When an error occurs the most likely consequence is that the robot will be in a legal configuration but within an unplanned environment, for example if a workpiece has been dropped or a component arrives unexpectedly at the cell. The error-recovery process may identify a new position for the robot so that it can participate in the recovery process. A new path must be planned which might include other active components such as a machine tool or a cooperating robot. The geometric model must support this re-planning exercise or at least decide it is intractable and refer to a higher authority.

The type of path planning to be attempted dictates the form in which the geometric model is stored. Brooks' method [4,5], already outlined, would require three sub-models to be maintained:

- A free-space model of the workpiece.
- A configuration-space model of the upper arm.
- A convenient object representation for forearm collisions.

Planning takes only a short time but may require significant computer power or a special architecture. A variation on the hypothesize-and-test method may be much more appropriate requiring only a conventional geometric model and relatively small computational resources at the expense of a risk that a particularly difficult path may not be found.

Some recent research [12] has tackled the problem of **error detection and recovery** in the presence of 'model error' which results from differences between a world model and the environment. We have so far considered strategies which are at best heuristic, while error-detection and recovery strategies are guaranteed to succeed if at all possible. Their 'axioms' are:

- A strategy should attain the goal when it is recognizably reachable and signal failure when it is not.
- It should permit speedy attainment of the goal.
- No motion guaranteed to terminate recognizably in the goal should ever be terminated prematurely.
- No motion should be terminated while there is any chance that it might achieve the goal due to unforeseen sensing and control events.

Error-detection and recovery strategies are devised in configuration space and are at an early stage of development so that they are not generally applicable. Successful application is in the long term.

In the short-to-medium term there will continue to be close operator integration with manufacturing cells. The role of human interaction in error recovery should be considered. The choice of robot trajectories in a recovery plan is one where a human operator could make suggestions and review the proposed alternatives. The level of optimization which needs to be performed by the real-time control system is then reduced, while the cell needs to present a human operator with a graphical display of the proposed recovery paths.

Handling time

During error recovery there will still be uncertainty in the time taken by various error-recovery processes. The uncertainty around relative timing is serious if there are several active cell components. As already discussed, the timing uncertainties need to be translated into geometric information which forms the basis for planning motions for recovery. For example consider an assembly cell in which one robot needs to perform a compliant motion while the other must move through free space. From experiments conducted during the cell design cycle we can deduce the likely extremely of timing in the two types of motion and then assess the 'worst case' close approach in the planned recovery motion.

When error recovery is being planned, access will be required to the timing of the various automatic processes. The timing of the recovery will need to be worked out using heuristics or classical tools (such as critical path analysis). The critical path of the recovery can be found and time-consuming processes identified. One technique identified for manufacturing systems is **temporal reasoning** [8].

A geometric criterion for errors

If uncertainties in the timing of a cell are reduced to their geometric impact, then we can consider a general framework for errors related to both timing and geometry.

A cell design and operating plan will be tolerant of a certain level of errors. For example a small misalignment in a part entering the cell may be accommodated by the gripper. A larger misalignment might be detected by a sensor and accommodated by a change in the robot approach trajectory; an excessive misalignment may cause a fault condition in which an external agent may need to be called.

In a similar way timing errors may occur on a scale of severity. Small timing errors will probably accommodate in the normal cell program; larger errors may need a slight change to the cell schedule to restore normal operation; severe timing errors may again cause a fault which needs a 'plan repair'.

Encoding the error recovery process

Referring again to the state model for the cell, an error can be considered to be an explicitly-defined error state or an undefined state. Error recovery must be achieved by a controlled series of state transitions. State transition can be represented in a number of ways,

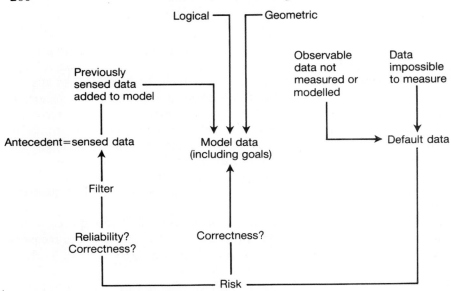

Fig. 14.15 Composition of a rule antecedent.

but one that is widely accepted is in the form of a **production** (one of a set of rules) which, if fulfilled, fires an appropriate state transition in the cell.

The rule consists of an antecedent which is to be fulfilled and a consequent which represents the action to bring about the state transition. The structure of the antecedent is shown in Fig. 14.15. The elements of the antecedent are sensed data which consists of raw and filtered data. Associated with this data is a risk, since the data may not be reliable; a sensor may have failed or the model on which the filter is based may be incorrect. Model data represents all the data that is assumed about the environment and what has been added since the cell was started up. Geometric data about the cell derived from the cell model is included in model data. Some aspects of the cell will not be modelled because they cannot be modelled or are uneconomic to model. For these some form of default will be required.

As error recovery proceeds and rules are analysed, the geometric data in the model database will be interrogated. This process suggests that the interface to the 'real-time' geometric model of the cell will consist of a large number of procedure calls which can be made from the rule evaluator. Consider a rule which calls for a return to the datum state (state 0). The geometric part of the antecedent will consist of a component which asks the model whether there is a path for the robot from where it is now to the datum state. The geometric modeller would then need:

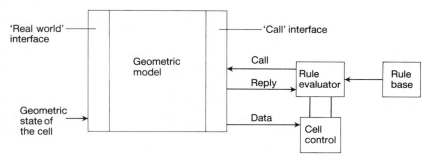

Fig. 14.16 Real-time geometric 'toolkit'.

- To confirm its own model by asking for the latest state (including joint angles, machine tool slide positions, gripper configuration) if it is not already recorded.
- To plan a path back to the datum position using knowledge of the cell free space.
- To check the path internally and if necessary to display the resulting path to the (human) cell supervisor.
- To return a 'true' result to the rule evaluator if a path is found, otherwise 'false' and to deposit the calculated trajectory in a return value area in the cell controller's memory.

Work is required to formulate the environment for the 'real-time' geometric modeller and the form of the call interface. The overall architecture of real-time cell-control systems is another research issue. The architecture of a real-time geometric 'toolkit' is shown in Fig. 14.16. There are two interfaces, one with the outside world and one with the internal parts of the error-recovery system. Data of two sorts are returned: Boolean values to the rule evaluator, and data pertinent to the operation of the cell to the cell controller.

14.4 Linking, planning and cell control

In the section on a geometric criterion for errors, we illustrated the need for geometric data in the error-recovery process. In this section we will consider the different requirements of off-line planning and real-time cell control.

In planning, we aim to produce some kind of specification for the cell operation. Taken overall, this will include programs for all the cell components as well as motion specification for the moving parts of the cell. Even if the cell operation is formally a plan, the 'action' parts of the plan will usually be in a conventional language, for example APT,

VAL or relay ladder logic. Where there must be a real-time resolution of the plan the 'action' element may be a process defined in the host language with symbolic parameters to be supplied at run-time. In a similar way contingency can be coded in this way.

In cell control the 'atomic' output of an error-recovery process may be a process with 'instantiated' variables but, where the error is unplanned, there may need to be a simple code generator within the cell controller. Extending this idea, the cell controller may even work in a 'universal' but simple control language and translation is done through a series of code generators appropriate to the cell hardware (cf. the CML universal machine control language [3]).

The cell controller will generally work with a simple geometric model and a correspondingly simple geometric modelling program. Apart from the considerations of computer resources which can be allocated to geometric modelling at the cell level, it is unlikely that the purchaser of the cell will either have or wish to purchase a full cell-modelling program based on a commercial solid modeller. Usually the systems house installing the cell will have the full cell-modelling tools while the user will have a 'slimmed-down' version of the modeller which will run in the cell's control system. The cell model will be simplified and down-loaded to the cell controller when necessary.

Object models

Our discussion of cell planning and control considered a number of model 'levels', each of which was appropriate to a stage of the planning process. For off-line planning we need access to the most sophisticated models possible so that we are free to do whatever approximations are required. In general the off-line model will be a solid model of either boundary or set-theoretic (CSG) type. Depending on the facilities offered by the particular geometric modeller we will be able to represent a range of surface types and gain access to the solid descriptions to make suitable approximations.

We suggested that a 'box' or wire-frame representation of the objects would probably be adequate for re-planning paths after an error had been identified. Both can be readily derived from solid models and are efficient on storage. The box representation offers a number of other advantages:

- It is easily derived from a boundary model by scanning its faces and edges. Refinement can be done by sectioning selected faces with planes.

- It is very efficient on storage requiring at most 24 real numbers to represent a cube as its 8 vertices.

- Box-to-box collision and distance checks can be done very quickly.

- The re-planning task will consist of a series of checks between bodies each represented by an array of boxes. The problem has a large degree of parallelism which might lend itself to a computer with a parallel architecture.

- Boxing is 'safe', in that it represents an upper bound on the space occupied by the object.

Either type of model will consist of two parts, the **body description** and the associated transformation. The body description remains the same during cell operation while the transformation changes in response to sensor and other information. When the error-recovery process is invoked, the transformations are applied to the body descriptions. The criterion of minimum interference with normal operation is achieved.

Straightforward boxing may not give sufficient accuracy. Invoking the 'multi-level' ideas already discussed, we may store the boxed representation at several levels of accuracy. Only at points of close approach will refinement be needed, and an extra level of boxing accuracy can be switched in depending on a computed minimum distance.

Resource planning

In the section on planning we discussed the problem of resources in the manufacturing cell. In general, resource allocation is resolved before run-time but where there is significant time variation in cell operation conflicts may arise. Resource allocation could be considered as a first level of error recovery in that it is modifying a previously computed plan to ensure continued operation.

Cell working space is the resource of primary interest and can be considered to be made up of:

- Space only ever occupied by one active device.
- Space occupied by more than one device at different times.

This spatial division would be best computed off-line where it could be analysed and approximated. The structure of representation is important for scheduling robot movement. For example we would wish to ensure that once one robot starts moving it does not have to come to a dead halt because the space required to finish the movement has already been allocated. Free-space representation should follow typical robot trajectories so that a scheduling algorithm can follow a sensible policy.

The run-time system would be required to retain the free space representation ready for scheduling. A simplified form would probably be appropriate but there would be trade-off between accuracy of representation and the quality of scheduling.

14.5 Conclusions

Path planning

We have reviewed path planning. Algorithmic techniques have been established but are exceedingly complex and time-consuming to compute. However they have demonstrated theoretical upper bounds for the problem solution. Free-space techniques are beginning to show results for practical problems but are still complex and need a lot of development before they can be generally applied. Such application to generalized cell planning is still in the long term. 'Semi-heuristic' techniques based on an analytical optimization of a suggested path represent a development step from the current practice of 'hypothesize and test'. Such techniques could be made fast and reliable. In the short-to-medium term such techniques can be implemented on current computer architectures.

Error recovery

We have seen that error recovery has a large number of elements of which geometry is just one. The challenge to computational geometry is to provide a real-time resource which can be used for error recovery in a general sense. We referred to a call interface which would be accessed by a rule evaluator, but such an interface would also be accessible from a procedural language.

In the short term a geometrical 'toolkit' could be developed to run on current architectures using:

- Simplified geometry.
- Semi-heuristic path planning.

In the medium term we might be able to improve the geometry and the path-planning techniques, while in the longer term the development of 'free-space' and error-detection and recovery strategies will improve the accuracy and resolution of the re-planning process.

Acknowledgements

We would like to thank Graham Jared of Cranfield Institute of Technology for his assistance with the BUILD geometric modelling program, Paul Rogers and Colin Dailly for their contribution to the research work on which this paper is based, and also Mike Duncan who prepared the machining cell models. This work has been supported by Cambridge Consultants Ltd., Trinity College Cambridge, KTM, Intel, Seescan and Cambridge Robotics Ltd.

References

1. E. W. Baumann, 'CAD model input for robotic sensory systems', Proceedings of AUTOFACT 4, Philadelphia, 1982.
2. M.C. Bonney, 'The simulation of industrial robot systems', *International Journal of Management Science* **12**,3, 1984.
3. D.A. Bourne, 'CML: A meta-interpreter for manufacturing', **private communication**, 1985.
4. R.A. Brooks, 'Planning collision free motions for pick-and-place operations', *International Journal of Robotics Research* **2**,4, 1983.
5. R.A. Brooks and T. Lozano-Pérez, 'A subdivision algorithm for findpath with rotation', Proceeding of the IJCAI, 1983.
6. S. Cameron, 'Modelling solids in motion', PhD Thesis, University of Edinburgh, 1984.
7. J. Carroll and D.J. Williams, 'Report of topic group on plan execution and monitoring', Alvey 5th Planning SIG Workshop, March 1986, **to be published**, IEE.
8. P. Cheeseman, 'A representation of time for automatic planning', IEEE International Conference on Robotics and Automation, 1984.
9. J.J. Craig, *Robotics Today*, 7 (45-47), 1985.
10. C. Dailly, PhD Thesis, University of Cambridge, 1986.
11. P. Davis 'Using and re-using partial plans', University of Illinois Co-ordinated Science Laboratory, Report R-772.
12. B.R. Donald, 'A theory of error detection and recovery: robot motion planning with uncertainty in the geometric models of the robot and the environment', NSF/SERC Workshop on Geometric Reasoning, Oxford, 1986.
13. H. Durrant-Whyte, 'Concerning uncertain geometry in robotics', NSF/SERC Workshop on Geometric Reasoning, Oxford, 1986.
14. B.R. Fox and K.G Kempf, 'A representation for opportunistic scheduling', Third International Symposium on Robotics Research, Paris, October 1985.

15. M. Gini and R.E. Smith, 'Reliable realtime robot operation employing intelligent forward recovery', University of Minnesota Institute of Technology Computer Science Department, Technical Report TR 85-30, September 1985.

16. J. E. Hopcroft, 'The impact of robotics on computer science', *Communications of the ACM* **29**, 1986.

17. J. Hopcroft and G. Wilfong, 'Motion of objects in contact', *International Journal of Robotics Research* **4**,4 (32-46), 1985.

18. M. L. Hornick and B. Ravani, 'Computer-aided offline planning and programming of robot motion', *International Journal of Robotics Research* **4**,4, 1986.

19. P. Howie, 'Graphic simulation for off-line robot programming', *Robotics Today*, February 1984.

20. O. Khatib, 'Real time obstacle avoidance for manipulators and mobile robots', IEE International Conference on Robotics and Automation, 1985.

21. M.H. Lee, N.W. Hardy and D.P. Barnes, 'Research into automatic error recovery', in UK Robotics Research 1984, Institution of Mechanical Engineers, London, December 1984.

22. T. Lozano-Pérez, 'Task planning', in *Robot Motion* (M. Brady et al., eds.), MIT Press, Cambridge, MA, 1984.

23. T. Lozano-Pérez and M. Wesley, 'An algorithm for planning collision-free paths among polyhedral obstacles', *Communications of the ACM* **22**, 1979.

24. O.Z. Maimon and S.Y. Nof, 'Co-ordination of robots sharing assembly tasks', *ASME Journal of Dynamic Systems, Measurement and Control* **107**, December 1985.

25. R.I. Mills, 'Computer simulation—a feasibility and planning tool for FMS', Proceedings of the 2nd International Conference on Flexible Manufacturing Systems, 1983.

26. I. Mitrani, *Simulation Techniques for Discrete Event Systems*, Cambridge University Press, 1982.

27. C. O'Dunlaing, M. Sharir and C. Yap, 'Generalized Voronoi diagrams for a ladder. I: topological analysis', Courant Institute Computer Science Department, Technical Report 139, 1984.

28. C. O'Dunlaing, M. Sharir and C. Yap, 'Generalised Voronoi diagrams for a ladder. II: efficient construction of the diagram', Courant Institute Computer Science Department, Technical Report 140, 1984.

29. C. O'Dunlaing and C. Yap, 'A "retraction" method for planning the motion of a disc', *Journal of Algorithms* **6**, 1985.

30. R.P. Paul, 'Sensors and the off-line programming of robots', Proceedings of the 8th ISR.

31. P. Rogers, 'A formalisation of the error recovery problem for manufacturing systems', CUED Manufacturing Group, International Report CUED/MG/IR/86/2, 1986.

32. K. Ruff, 'Contemporary manufacturing systems integration', University of Michigan NSF Workshop on Systems Integration Tools in Manufacturing, November 1985.

33. J.T. Schwartz and M. Sharir, 'On the piano movers' problem. I: the case of a two-dimensional rigid polygonal body moving admidst polygonal barriers', *Communications on Pure and Applied Mathematics* **36**, 1983.

34. J.T. Schwartz and M. Sharir, 'On the piano movers' problem. III: coordinating the motion of several independent bodies: the special case of circular bodies moving amidst polygonal barriers', *Robotics Research* **2**,3, 1983.

35. J.T. Schwartz and M. Scharir, 'A Survey of motion planning and related geometric algorithms', NSF/SERC Workshop on Geometric Reasoning, Oxford, 1986.

36. M. Sharir and E. Ariel-Sheffi, 'On the piano movers' problem. IV: various decomposable two-dimensional motion planning problems', *Communications on Pure and Applied Mathematics* **37**, 1984.

37. R.N. Stauffer, 'Robots 9: signs of a maturing industry', *Robotics Today*, August 1985.

38. R.K. Stobart, 'Using solid modellers in robot programming', Proceeding of the BRA 8th Annual Conference, Birmingham, 1985.

39. R.K. Stobart, 'Collision detection for the off-line programming of robots', IFIP Conference on Off-line Programming of Industrial Robots, Stuttgart, 1986.

40. R.K. Stobart and C. Dailly, 'The use of simulation in the off-line programming of robots', IEE Seminar on UK Robotics Research, 1985.

41. S.M. Udupa, 'Collision detection and avoidance in computer controlled manipulators', PhD Dissertation, California Institute of Technology, Pasadena, 1977.

42. V. Vemurs, *Modelling of Complex Systems: An Introduction*, Academic Press, New York, 1978.

43. M. Weck, T. Niehaus and M. Osterwinter, 'An interactive model-based robot programming and simulation workstation', IFIP Conference on Off-line Programming of Industrial Robots, Stuttgart 1986.

44. C. Widdoes, 'A heuristic collision avoider for the Stanford robot arm', Stanford AI Laboratory Report.

45. B. Yin, 'A framework for handling vision data in an object level robot language—RAPT', University of Edinburgh Department of Artificial Intelligence Research, Paper 189.

Discussion

(Paper presented by D. Williams)

Bowyer: I was at a conference a few months ago where somebody said that potential methods for path planning were a black hole for engineers! They lock into their mental receptors like an encephalin—they never do anything else once they've heard about them. The problem, of course, is that potential methods get stuck at a local minimum. Has anyone that you know of been doing work on using simulated annealing in potential path planning?

Dodsworth: I have a Ph.D. student (C. Lindsay) who has been doing just that. He is writing up at the moment.

Cameron: Going back to your picture of configuration space for a cylindrical robot, there are a couple of things that you can add to that. Firstly, you can put in obstacles in configuration space which aren't due to obstacles in the world: for instance, joint limits. Secondly, with your arm, it looked like you needed another region for when the arm was close in, and the back of the arm was then pointing outwards.

D. Williams: Right: I didn't mean that to be a very rigorous analysis of that configuration space; it was just an illustration.

Cameron: And, going back to the potential-field stuff, somebody called Charles Buckley has just been working on a potential-field method based on a different potential-field function [46]. I still haven't worked out whether it's much good yet.

D. Williams: My work is based on what I've read of Osama Katib's work at Stanford [20].

Cameron: This was done at Stamford too, as a Ph.D.

Additional reference

46.　　C.E. Buckley and L.J. Leifer, 'Proximity metric for continuum path planning', Proceedings of the 9th International Joint Conference on Artificial Intelligence, Los Angeles (1096-1102), 1985.

15 Using computer-aided design and expert systems for human workplace design

M. C. BONNEY, N. K. TAYLOR and K. CASE

15.1 Introduction

The prime objective of a human workplace design system is to produce designs which are 'suitable' for the user population. Suitability is taken to include functionality, cost, reliability and appearance. To achieve this it is essential to bring together the designer's expertise in workplace design, with knowledge of the task and of the potential user.

The traditional manual design process is a practical way to produce workplace designs. Another way is for designers/ergonomists to work in conjunction with computer-aided design software systems such as SAMMIE. Other approaches are being explored. For example, ALFIE is an expert system which is able to address some ergonomic problems associated with workplace design.

This paper examines how 'traditional' computer-aided design, expert systems, geometric reasoning and designers may be brought together to produce an effective tool for human workplace design. It examines first the requirements of a general human workplace design system. It then summarizes features of the SAMMIE ergonomic design system, looks at the ALFIE expert systems and suggests how SAMMIE and ALFIE may be combined, possibly with geometric reasoning and other facilities, to produce a total design package.

15.2 The requirements of a general workplace design system

In the broadest sense human 'workplaces' include all physical environments that humans find themselves in when at work, taking part in recreational activities, domestic situations, etc. Hence a human workplace design system needs to be suitable for a wide range of application areas. The overall objective of a computer-based system should be to enhance the flair of the designer by the addition of the powerful resources of the computer. This immediately raises the question of the allocation of design functions between designer and computer. A traditional aid to resolving this question has been the use of Fitt's lists [8]. These are lists of the relative abilities of human and machine based on a range of performance criteria, the idea being that some composite collection of individual attributes will be suitable for the overall system requirements. The putting together of an 'identikit' of 'good' parts of the person (designer) and machine (computer-aided design system) is unlikely to result in a good overall system. Instead we should be looking for systems where designer and computer work on aspects of the problem in a mutual rather than exclusive fashion. The changing characteristics of computer hardware and software (including expert systems and, now, geometric reasoning) not only alter the balance between human and machine capabilities, but also have potential for providing the essential mechanism for the desired synergistic relationship.

The creation and evaluation of design alternatives is an important attribute of any design system, and computer-aided design systems have frequently claimed ability in this area. In the context of workplace design, evaluations are required of the physical layout of the equipment, the suitability of the workplace geometry and environment for the user, the requirements of the task and other factors.

Evaluations of the physical layout of the equipment include whether the layout aids the flow of materials and information and geometric considerations of layout, such as whether the items will fit. The suitability of the workplace geometry requires checking that the user can reach the appropriate locations, see the relevant places, take up suitable postures and apply appropriate forces. The workplace also needs to be suitable for humans with regard to environmental factors such as temperatures, light intensity, noise levels, humidity and vibration. These environmental criteria are normally dependent on the task; important considerations are mental and physical loads and the time required to perform it.

Alongside the above there are many other factors common to all design situations including life-cycle costing, ease of manufacture, marketability, customer appeal, reliability and maintainability. The final workplace design will also need to undergo user trials, and there is a necessity for the design system itself to be usable and cost effective.

15.3 SAMMIE—a computer-aided ergonomic design system

Simulation reduces the necessity to build physical mock-ups for user trials and improves the quality and usefulness of any prototypes that are built. In this way it is expected that the increasing pressures from social, economic, and legislative forces for good ergonomic design can be more easily accommodated within the functional design process. SAMMIE (System for Aiding Man-Machine Interaction Evaluation) provides a computer-based three-dimensional building and viewing scheme for modelling, together with an anthropometrically and biometrically variable three-dimensional model of the human body. A set of human capability evaluative techniques, based on the model and including reach, vision and fit, are available and the complete system is accessed through a user interface which assists in the interactive nature of the design and evaluation process.

The workplace modelling system is used to generate geometric representations of the workplace and specific items of equipment. A relatively simple form of solid model, the **boundary model**, is used as it is both suitable for the application and provides an adequate response to interactive changes in design. Solid shapes are defined by the specification of vertices, edges and plane polygon faces, but in many instances the geometry specification is eased by the use of a set of simple parametrically defined shapes such as cuboids and prisms. Relatively complex models can be built and swiftly manipulated to change the geometry or to change the view as seen on the screen.

The SAMMIE modeller makes use of a hierarchical data structure to specify spatial relationships between geometric items in the model in the logical or functional manner. Thus the opening of a door of a car model is a meaningful operation; should the model contain a door handle then it would be interpreted as a sub-part of the door and would maintain the appropriate spatial relationship as the door is opened. The use of hierarchical data structures in this way is strongly developed within SAMMIE as an essential evaluative tool and is implemented with a suitable user interface. Interactive modification of geometry in ways relevant to the design situation is also an important part of the system. For example, if a table were modelled as a table top and four legs, then increasing the length of the top would automatically re-position the legs to maintain a valid model.

A wide variety of viewing options are available including orthographic projections, perspective, viewing point and centre of interest control, scaling, etc. In addition to being part of the user interface enabling better comprehension of the model, viewing is also available as a model evaluation facility in its own right. Hence the view

SOMATOTYPES

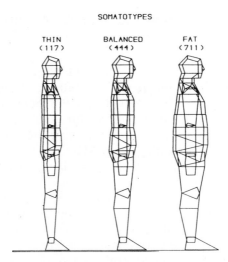

Fig. 15.1   Examples of the flexibility of the SAMMIE man model.

as would be seen by the human operator model from within the geometric model can be presented, as can special views such as those seen in mirrors.   The facilities have been further extended so that two-dimensional visibility plots and three-dimensional visibility charts can automatically be generated and displayed [8].

The **man model** provides much of the evaluative power of SAMMIE through its anthropometric and biomechanical modelling capabilities. It consists of a set of pin joints and straight rigid links structured hierarchically to represent the major points of articulation and the body segment dimensions.   A three-dimensional flesh shape based on a somatotype classification [10] is arranged about this link structure using the modelling methods described above. This is illustrated in Fig. 15.1. Parameters for each of the body segments are extracted from a database which contains definitions of the relevant population.

Interactive techniques enable the anthropometry to be rapidly amended by changing the overall body percentile, an individual link percentile, or an explicit link dimension. Correlation equations are used to relate internal dimensions to externally measurable dimensions. Clearly the selection of an adequate database and the subsequent manipulation of the information requires a thorough understanding of the anthropometric implications.   The SAMMIE software provides a user interface for varying the anthropometry of the man model.

Several methods are available for the manipulation of the man model within the workplace model. Postures can be created and stored within the database to be subsequently recalled. These postures can simply be a set of potentially useful starting points for the investigation of actual working postures or they can be a precise set against which potential designs must be evaluated. Each body segment can be articulated about its proximal joint. Movement is permitted in the flexion-extension, abduction-adduction and medial-lateral rotation senses and the resulting joint posture is compared with joint constraints in the database. The system reports whether the joint is within 'normal' range of movement, within the maximum, or infeasible. The 'normal' constraint data permits the user to define design criteria in terms of body posture. Hence, for example, if high gravity forces precluded any attempt to raise the arms above shoulder height then this could be accommodated in the database. In more usual conditions this facility is used to define preferred working volumes related to a joint postural comfort criterion.

Reach algorithms are available which predict a feasible posture for a sequential set of links such as the arm or leg. In the evaluation situation, the ability to test reach to the specific points, where for example controls are located, is a useful facility. However in a design situation it may be necessary to determine suitable areas or volumes within which controls could be placed and for this application 'reach contours' have been developed. These enable envelopes of areas within reach to be overlaid on any surfaces of the model as an aid to assessing suitable positions for control locations. A major study involving this facility determined reach zones for the drivers of agricultural tractors [9]. Some of the problems of reach and posture associated with this field of study are illustrated in Fig. 15.2.

Further information on SAMMIE is available in the literature [2,6,7].

15.4 ALFIE—an expert aid for ergonomic design

In recent years an expert aid for ergonomic design called ALFIE (Auxiliary Logistics For Industrial Engineers) has been constructed [11]. As the acronym implies, the system has a more general applicability than the domain of ergonomics. It is expected to find applications in other areas of industrial design since the problems faced in ergonomic design are by no means unique. The underlying problems solving processes employed in design-type tasks appear to have a considerable degree of homogeneity.

When used as a ergonomic design tool ALFIE is intended to be used by industrial designers and design engineers who have no great knowledge of the field of ergonomics. The general aims of the system are:

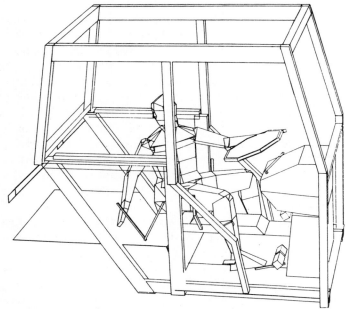

Fig. 15.2 A typical example of a SAMMIE application.

- To guide the users to important ergonomic concepts.

- To invoke appropriate ergonomic rules and models and thereby update design parameters.

- To highlight the relationships between the design parameters.

- To assist the user in constraining the design parameters.

- To advise the user of any problems which may arise.

- To provide a summary of the ergonomic constraints on the design parameters which should be used to constrain the overall design.

It is in the nature of design that it is an interactive process broken up by conceptual leaps, so it is essential that any design aid does not thwart the creative potential of the more experience designer. In systems where the design options are many the creativity of the user is of the utmost importance in pruning the search space of design possibilities. Ergonomic design falls into this category.   There are no clear optimization procedures and a dearth of constraints compared to the numerous options available. The dual aims of ALFIE are, therefore, to provide the ergonomic novice with the detailed guidance required and

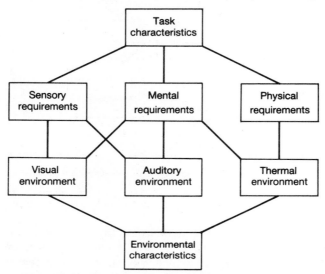

Fig. 15.3 An example of an ALFIE concept net.

yet allow the more experienced user the autonomy to move freely around the system examining the areas desired.

The system has been designed to handle both logical and numerical formulations of knowledge and combines a rule-based with a model-based approach. In addition, the system permits considerable vagueness on the part of the user by not insisting on fixed values for quantifiable design parameters. The user is allowed to give ranges for these values but is expected to 'tighten up' the ranges as he settles on the details of the design.

In order to break down the domain of ergonomics into manageable sub-fields the system contains a network of 'concepts'. Each concept represents some body of ergonomic knowledge and is linked, throughout the network, to other relevant areas of ergonomics. This is illustrated in Fig. 15.3. Apart from assisting the knowledge engineer to visualize the system, these concepts permit the user to go straight to those areas which are considered important without being forced through a chain of rules, some of which might be totally irrelevant to the current application. The ALFIE user obtains design assistance by investigating ergonomic concepts which are of interest to him. A guidance system assists the user in his investigations if he is uncertain about which concepts are relevant. The investigation of a concept is achieved by firing a number of 'rules' which are associated with that concept. These rules lead to the provision of advice, the investigation of other concepts, the firing of other rules and the modification of design parameters.

The parameters of a design are represented by 'factors' in the ALFIE system. Factors can be updated automatically or by requesting values

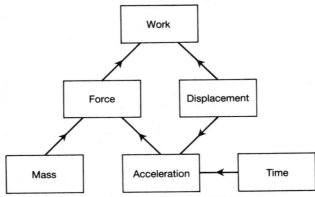

Fig. 15.4 An example of an ALFIE influence net.

from the user. They form the basis of the conditions in the rules. Factors are also linked together, as in Fig. 15.4, to form an influence net which indicates the effect which a change in each factor will have on any of the others. The influence net is built by the system from an analysis of the specified relationships between the factors. The relationships are basically equations and are presented by 'models' within the ALFIE system. These models also permit the automatic evaluation of factors.

The system has been employed successfully to examine heat stress problems and strength problems, both of which utilize mainly numerically formulated knowledge and so require a model-based approach. It has also been used to build knowledge bases on illumination levels and inspection problems which, due to the large number of propositions involved in the knowledge formalism, required a rule-based approach.

15.5 Linking SAMMIE and ALFIE

The human designer uses SAMMIE to answer certain design questions such as whether the human operator can reach to specified points. ALFIE, on the other hand, is largely rule-based and by question and answer the designer is given guidance on the suitability of the values of certain design parameters. Fig. 15.5 illustrates in schematic form how SAMMIE and ALFIE interrelate. Our present plans are that they should be joined together by means of common applications data and, more particularly, by appropriate processing of data so that the two systems assist each other. An example of this is the evaluation of heat stress of a pilot in a cockpit which is subject to solar gain through the canopy. SAMMIE would enable the workplace to be modelled and

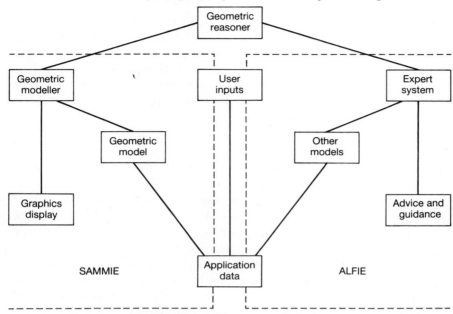

Fig. 15.5 Proposed integration of the SAMMIE and ALFIE systems.

could be adapted to calculate information related to heat gain. This could then be passed across to ALFIE and used as part of the design process to provide guidance on the use of cooling systems, tinted glass etc. The effect of different cockpit geometry—overall dimensions and size of particular window sections—would also affect the heat gain.

One area in which SAMMIE and ALFIE could be combined fruitfully is in the design of workplaces where visibility is of significant importance. Currently, SAMMIE visually aids the designer of such a workplace by presenting the man's direct and mirror views. The designer studies these and applies his own set of rules to modify the workplace in order to arrive at an acceptable solution. These rules could be applied equally well within the ALFIE system. However, in order to apply them the ALFIE system would need to be given the ability to 'see' what the SAMMIE system is showing. Both ALFIE and SAMMIE are computer-based, and so the real problem is that of passing to ALFIE information which the designer gains by seeing the SAMMIE display. This information includes the size of the field of view and whether certain items are visible. Both could be derived by a suitable set of algorithms which form the geometric reasoning required to link SAMMIE and ALFIE as illustrated in Fig. 15.5.

Giving the user of SAMMIE access to ALFIE would provide him with the ergonomic knowledge which he needs. In addition to the savings provided by this on-line expertise, the integration of the

SAMMIE data structure with the ALFIE knowledge base reduces the amount of information transfer required. Access to SAMMIE would provide the user of ALFIE with a geometric model of the design being developed or evaluated. The link required for this need be no more than a shared application database. However, there is scope for an even greater benefit by removing some of the spatial and/or temporal reasoning required of the designer.

15.6 Geometric reasoning and other needs

We are using *geometric reasoning* to mean the application of computer techniques to spatial problems so that deductions can be made from the geometry. Our aim is to apply geometric reasoning to geometrically complex domains in a multivariate decision environment.

The location and dimensions of the workplace will affect the performance of the operator and thus many of the decisions facing a workplace designer are geometric. The problem is to select appropriate geometric measures to guide good ergonomic design.

It is intended that SAMMIE will be the front end to a combined SAMMIE/ALFIE system and that the hierarchial data structure within SAMMIE will provide a framework which will aid geometric reasoning. The geometric reasoner will provide ALFIE with some sort of 'spatial awareness'. Furthermore, geometric reasoning enhances SAMMIE by allowing rule-based interpretations of the geometry.

ALFIE needs information in the form of distance between objects. Location predicates such as 'above', 'on top of', 'below' and more complex application predicates such as 'in view', 'obscured', 'in reach' are also needed. SAMMIE uses a geometric modeller which deals with information at a much lower level—coordinates, vertices, edges and faces. To go from the coordinates of two objects to a predicate of the form 'above' requires geometric reasoning. 'In view' and 'obscured' require geometric reasoning of greater power.

It is worth noting that a temporal reasoner would also be useful to ALFIE. By using time predicates such as 'before', 'after', 'simultaneously' and information in the form of delays and free time between events, ALFIE would be able to access any temporal knowledge within the computer-aided design system.

This paper examines ergonomic decisions which arise in workplace design and which are geometrically related. Some of these decisions are of a fairly general nature whereas others are more applications-specific and may interact with functional needs.

Examples of ergonomic geometric design facilities are the assessment of posture, forces, comfort, energy expenditure, spinal loading, etc.

These would require the system to have knowledge of gravitation, masses, position and time. Another example is the identification of interference between the user and the workplace, for example to ensure that protuberances are not at certain heights. At a detailed level, location of guards for machinery may require appropriate clearances. Some examples of applications-specific design facilities are:

- When assessing visibility in rear view mirrors where is a need for knowledge of EEC legislative requirements.

- When designing coal face equipment there is a need to consider interactions between the user and legislation on equipment and safety in mines.

15.7 Concluding remarks

Typical ergonomic design problems cover such diverse issues as lighting levels, temperatures, noise levels, anthropometry, posture, comfort, reach, physical strength, visibility and safety legislation.

In order to evaluate these issues and their interactions in a satisfactory manner a knowledge of the workplace geometry often needs to be combined with a knowledge of desirable and legal requirements. Questions which need answering include:

- Are the noise levels within legal requirements? This requires that, for a given workstation, the relative locations of the noise sources and any absorbing surfaces must be known in order to derive the noise level. This can then be compared with the legal limits and modifications made to the workplace layout if necessary.

- Can the operator reach the controls? Again, the locations of the controls relative to the operator are required together with anthropometric information in order to answer this question.

- How long will the operator be able to perform the task? This question draws on information of physical strength, temperatures, distance moves, speeds required and a host of other factors.

Clearly a knowledge of the geometry of the workplace is essential to the solution of a large number or ergonomic problems. The ability to reason about this geometry is therefore a prerequisite for a truly comprehensive design aid for ergonomics. An extension of the geometric reasoner to include the fourth dimension—time—would permit the combination of knowledge about endurance, dynamic force factors and required work rates to derive even better estimates of work capacity, work/rest, etc.

Some of the problems discussed in this paper have already been faced by the researchers at the University of Nottingham during the development of a number of computer systems constructed to meet the specific requirements of particular workplace categories.

For instance, the AUTOMAT (Automatic Methods and Times) computer-aided work-measurement system [4] produces workplace layouts suitable for tasks which are specified in a high-level language. The layout is based on heuristic rules which locate items for appropriate use of tools, balance work between the hands and place heavy items within the two-handed work area. An analysis of methods and times for the tasks are derived by the computer using the task description, the location of the operator and the parts, the handling characteristics of the parts and their relationship to the tooling.

CAPABLE (Control and Panel Arrangements by Logical Evaluation) [5] is, as its name implies, a system directed to the problems of control and panel layouts. It uses heuristics based on a range of factors such as control separation, work distribution between limbs, distance, time, comfort, percentiles, frequency of operation and importance. These heuristics are used to derive control and panel arrangements for an operator specified in position and dimensions so as to optimize an objective function which is a weighted linear combination of the measures. A high-level task language is again used and time and geometric assessments, such as separation of controls, are used in conjunction with rules.

GRASP (Graphic Robot Applications Simulation Package) [3,5] is concerned with the design and evaluation of AMT work cells. The origins of GRASP lay in SAMMIE which was restructured to meet the needs of robot workplace design problems. It is time-coordinated, and contains an event processor which allows it to simulate the action of objects and equipment in addition to robots. There are high-level robot programming facilities, a kinematic modeller and tool path control. The output from the system can be post-processed to become a robot program. Signalling facilities can also be simulated and permit the specification of actions which are conditional on sensory inputs and feedback. GRASP illustrates some of the facilities which would be required for a temporal reasoner.

There are many potential benefits to be derived from the marriage of the fields of computer-aided design and intelligent knowledge-based systems. Geometric reasoners provide a vehicle by which this could be achieved and we are investigating the feasibility of such a development in the area of human workplace design.

Acknowledgements

The authors would like to thank the Science and Engineering Research Council which funded much of the work described and to acknowledge the help received from the British Technology Group.

References

1. M.C. Bonney et al., 'The simulation of industrial robot systems', *OMEGA International Journal of Management Science* **12**,3 (273-281), 1974.

2. M.C. Bonney, C.A. Blunsden, K. Case and J.M. Porter, 'Man-machine interaction in work systems', *International Journal of Production Research* **17**,6 (619-629), 1979.

3. M.C. Bonney, R.G. Marshall and J.L. Green, 'Off-line programming using the GRASP simulator', *IFIP Working Conference on Off-line Programming of Industrial Robots*, Stuttgart, 1986.

4. M.C. Bonney and N.A. Schofield, 'Computerised work study using the SAMMIE/AUTOMAT system', *International Journal of Production Research* **19** (321-336), 1971.

5. M.C. Bonney and R.W. Williams, 'CAPABLE: a computer program to layout controls and panels', *Ergonomics* **20**,2 (297-316), 1977.

6. K. Case, J.M. Porter and M.C. Bonney, 'SAMMIE: A computer-aided design tool for ergonomists', Proceedings of the 12th Human Factors Meeting, Dayton, 1986.

7. K. Case, J.M. Porter, M.C. Bonney and J. Levis, 'Design of mirror systems for commercial vehicles', *Applied Ergonomics* **11**,4, 1980.

8. P.M. Pitts et al, *Human Engineering for an Effective Air Navigation and Traffic Control System*, Washington DC National Research Council, 1951.

9. J.M. Porter, K. Case and M.C. Bonney, 'Computer generated three-dimensional visibility chart', *Human Factors in Transport Research* (D.J. Oborne and J.A. Levis, eds.), Academic Press (Vol 1: 265-373), 1980.

10. C.J. Reid, S.A. Gibson, M.C. Bonney and D. Bottoms, 'Computer simulation of reach zones for the agricultural driver', Proceedings of the 9th International Congress of International Ergonomics Association, Taylor and Francis, 1985.

11. W.H. Sheldon, *The Varieties of Human Physique*, Harper, New York, 1940.

12. N.K. Taylor and E.N. Corlett, 'ALFIE—auxiliary logistics for industrial engineers', *International Journal of Industrial Ergonomics*, **in press for 1987.**

Discussion

(This paper was presented by Bonney)

Irvine: Could I ask a question about using SAMMIE. An awful lot of ergonomics consists of remedial ergonomics, where an ergonomist is brought in when a design is at a very late stage, or even, when it is finished and is in operation, and he is trying to improve the thing from the point of view of the worker. Have you used SAMMIE at all when you have all these terrible restrictions on what you can actually do after you've been brought into the setup?

Bonnie: We have used the system for about thirty industrial applications, and most of them were already highly constrained, but I think that it's much more valuable if you can jump in without those limitations. I think that ergonomists are much more open to this now; they want to come in early. A lot of the problem has been that design procedures often excluded the ergonomist until very late in the day. Then they said, "let's make it very nice for people", by which time there was inadequate space, or whatever. For example, we got involved with a fork-lift truck where the driver couldn't see the forks. They had already made a prototype, and were some way down the path. We said that there was no way that it was usable. Twenty weeks later, when they got the prototype back and evaluated it, they found that out. And we were able to offer some constructive suggestions as to what they could move to make the forks visible.

Cameron: More a comment than a question: we have had some thoughts at Edinburgh about this idea of using a geometric modeller as a black box. We actually went ahead and connected the ROBMOD modeller that we had up there to the RAPT system. I've just published a paper on that, at a NATO Conference published by Springer Verlag, which will be coming out shortly [13].

Middleditch: Could we finish this meeting with a definition of geometric reasoning?

Woodwark: Didn't the last speaker give one?

Todd: Surely, the variety of papers presented at this meeting shows that a definition of geometric reasoning is not possible.

Bowyer: The more useful that something is, the less easy it is to define!

Additional reference

13. A.P. Ambler, S.A. Cameron and D.F. Corner, 'Augmenting the RAPT robot language', in *Languages for Sensor-Based Control in Robotics* (U. Rembold and K. Hormann, eds.) (Proceedings of a NATO ASI, Castelveccio Pascali, Italy, 1986) (305-316), 1987.